Collins

Cambridge Lower Secondary

Maths

STAGE 7: WORKBOOK

Alastair Duncombe, Rob Ellis, Amanda George,
Claire Powis, Brian Speed
Series Editor: Alastair Duncombe

Collins

William Collins' dream of knowledge for all began with the publication of his first book in 1819. A self-educated mill worker, he not only enriched millions of lives, but also founded a flourishing publishing house. Today, staying true to this spirit, Collins books are packed with inspiration, innovation and practical expertise. They place you at the centre of a world of possibility and give you exactly what you need to explore it.

Collins. Freedom to teach.

Published by Collins
An imprint of HarperCollins*Publishers*
The News Building
1 London Bridge Street
London
SE1 9GF

HarperCollins*Publishers*
Macken House, 39/40 Mayor Street Upper,
Dublin 1, D01 C9W8, Ireland

Browse the complete Collins catalogue at
www.collins.co.uk

MIX
Paper | Supporting responsible forestry
FSC™ C007454

This book contains FSC™ certified paper and other controlled sources to ensure responsible forest management.

For more information visit: www.harpercollins.co.uk/green

British Library Cataloguing in Publication Data
A catalogue record for this publication is available from the British Library.

Authors: Alastair Duncombe, Rob Ellis, Amanda George, Claire Powis, Peter Ransom, Brian Speed
Series editor: Alastair Duncombe
Publisher: Elaine Higgleton
In-house project editors: Jennifer Hall and Caroline Green
Project manager: Wendy Alderton
Development editors: Anna Cox, Rachel Hamar and Phil Gallagher
Copyeditor: Alison Bewsher
Proofreader: Tim Jackson
Answer checker: Eric Pradel and Jouve India Private Limited
Cover designer: Ken Vail Graphic Design and Gordon MacGlip
Cover illustrator: Ann Paganuzzi
Typesetter: Jouve India Private Limited
Production controller: Lyndsey Rogers
Printed in India by Multivista Global Pvt. Ltd.

Acknowledgements

The publishers gratefully acknowledge the permission granted to reproduce the copyright material in this book. Every effort has been made to trace copyright holders and to obtain their permission for the use of copyright material. The publishers will gladly receive any information enabling them to rectify any error or omission at the first opportunity.

p.29 Victor Brave/Shutterstock; p.133 United States Environmental Protection Agency; p.134 Marmotte Granfondo Series; p.135 Contains public sector information licensed under the Open Government Licence v3.0.; p.135 Contains public sector information licensed under the Open Government Licence v3.0.

Cambridge International copyright material in this publication is reproduced under licence and remains the intellectual property of Cambridge Assessment International Education.

Third-party websites and resources referred to in this publication have not been endorsed by Cambridge Assessment International Education.

With thanks to the following teachers and schools for reviewing materials in development: Samitava Mukherjee and Debjani Sen, Calcutta International School; Hawar International School; Adrienne Leisztinger, International School of Budapest; Sujatha Raghavan, Manthan International School; Mahesh Punjabi, Podar International School; Taman Rama Intercultural School; Utpal Sanghvi International School.

Contents

How to use this book

This Workbook accompanies the *Collins Lower Secondary Maths Stage 7 Student's Book* and covers Cambridge Lower Secondary Mathematics curriculum framework (0862). This Workbook can be used in the classroom or as homework. Answers are provided in the Teacher's Guide.

Every chapter has these helpful features:

- 'Summary of key points': to remind you of the mathematical concepts from the corresponding section in the Student's Book.

- Exercises: to give you further practice at answering questions on each topic covered in the Student's Book. The questions at the end of each exercise will be harder to stretch you.

- 'Thinking and working mathematically' questions (marked as   ): to help you develop your mathematical thinking. The questions will often be more open-ended in nature.

- 'Think about' questions: encourage you to think deeply and problem solve.

1 Factors

You will practice how to:

- Use knowledge of tests of divisibility to find factors of numbers greater than 100.
- Understand lowest common multiple and highest common factor (numbers less than 100).

1.1 Divisibility tests

Summary of key points

A number is **divisible** by another number if it can be divided exactly by that number without leaving a remainder. For example, 12 is divisible by 6 because 12 ÷ 6 = 2 with no remainder.

A **divisibility test** is a quick method for checking whether one number is divisible by another number.

A number is:

divisible by 2 if it is even

divisible by 3 if the sum of its digits is divisible by 3

divisible by 4 if the number made by its last two digits is divisible by 4

divisible by 5 if its last digit is 0 or 5

divisible by 6 if it is even and divisible by 3

divisible by 8 if the number made by its last three digits is divisible by 8

divisible by 9 if the sum of its digits is divisible by 9

divisible by 10 if its last digit is 0

divisible by 25 if its last two digits are 00, 25, 50 or 75

divisible by 50 if its last two digits are 00 or 50

divisible by 100 if its last two digits are 00.

Exercise 1

1. Here is a list of numbers: **135, 140, 144, 149, 155, 162, 170, 179.**

   a) Write the numbers from the list that are divisible by 2.

   ....

   b) Write the numbers from the list that are divisible by 3.

   ....

   c) Write the numbers from the list that are divisible by 4.

   ....

d) Write the numbers from the list that are divisible by 6.

....

e) Write the numbers from the list that are divisible by 9.

....

2 Here is a list of numbers: 1512, 3870, 4220, 5000, 7065, 8226.

a) Write the numbers from the list that are divisible by 4.

....

b) Write the numbers from the list that are divisible by 6.

....

c) Write the numbers from the list that are divisible by 8.

....

d) Write the numbers from the list that are divisible by 9.

....

3 Here is a list of numbers: 45 072, 46 100, 47 310, 48 501, 49 816.

a) Which number is divisible by 25?

....

b) Show that 45 072 is divisible by 6.

....

....

c) Write the numbers from the list that are divisible by 8.

....

d) Write the numbers from the list that are divisible by 9.

....

4 The number 715 ?34 is divisible by 9. What is the missing digit? ....

5 Find the number between 3158 and 3164 that is divisible by 6. ....

6 Find the largest number less than 50 000 that is:

a) divisible by 100 .... **b)** divisible by 6 ....

c) divisible by 3 ....

7 Find the smallest number greater than 77 050 that is:

    a) divisible by 5 ....

    b) divisible by 9 ....

    c) divisible by 8 ....

Think about

8 Find all the factors of 300. Think about what strategies to use.

1.2 Lowest common multiple and highest common factor

Summary of key points

The **lowest common multiple (LCM)** of two or more whole numbers is the lowest multiple shared by the numbers. For example, 4 and 5 have common multiples of 20, 40, 60, etc. The lowest common multiple of 4 and 5 is 20.

The **highest common factor (HCF)** of two or more whole numbers is the highest factor shared by the numbers. For example, 16 and 24 have common factors 1, 2, 4 and 8. The highest common factor of 16 and 24 is 8.

Exercise 2

1 Write the highest common factor of:

    a) 9 and 18 ....
    b) 24 and 32 ....

    c) 60 and 80 ....
    d) 55 and 70 ....

    e) 18, 24 and 30 ....
    f) 30, 45 and 75 ....

2 Write the lowest common multiple of:

    a) 9 and 18 ....
    b) 10 and 11 ....

    c) 14 and 35 ....
    d) 20 and 50 ....

    e) 6, 9 and 15 ....
    f) 12, 16 and 24 ....

3 Write two numbers with highest common factor:

    a) 4 ....
    b) 10 ....

4 Without writing the number 1, write two numbers with lowest common multiple:

a) 14   ....

b) 20   ....

c) 36   ....

5 Olga and Carlos each buy some packets of pens. The packets all contain the same numbers of pens.

Olga buys 84 pens.

Carlos buys 48 pens.

What is the largest possible number of pens in each packet?....................

6 The children from Valley Primary School are going on a school trip. They can all travel on buses with 18 seats, with no seats empty. Alternatively, they can all travel on buses with 24 seats, with no seats empty. What is the smallest possible number of students in the school?

....

7 A violin competition takes place every 3 years. A piano competition takes place every 8 years.

Both competitions took place in 2019.

When is the next year when both competitions will take place?

....

Think about

8 p and q are prime numbers.

a) Write the highest common factor of p and q.

b) Write the lowest common multiple of p and q.

2 2D and 3D shapes

You will practice how to:

- Understand that if two 2D shapes are congruent, corresponding sides and angles are equal.
- Know the parts of a circle:
  - o centre
  - o radius
  - o diameter
  - o circumference
  - o chord
  - o tangent
- Identify and describe the combination of properties that determine a specific 3D shape.

2.1 Congruency

Summary of key points

Two shapes are **congruent** if they have exactly the same shape and size.

Corresponding sides and angles of congruent triangles are equal.

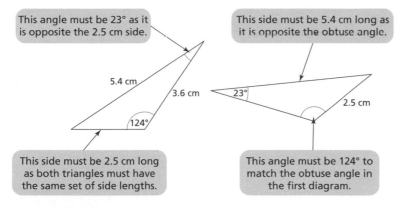

This angle must be 23° as it is opposite the 2.5 cm side.

This side must be 5.4 cm long as it is opposite the obtuse angle.

This side must be 2.5 cm long as both triangles must have the same set of side lengths.

This angle must be 124° to match the obtuse angle in the first diagram.

Exercise 1

1 The diagram shows two congruent isosceles triangles.
Mark on the missing measurements.

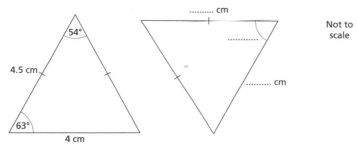

Not to scale

2 Triangles *ABC* and *PQR* are congruent.

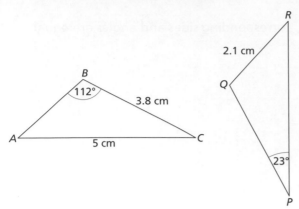

Not to scale

Are these statements true or false?

	True	False
a) Angle *BAC* = 23°	☐	☐
b) *PQ* = 5 cm	☐	☐
c) Angle *RQP* = 112°	☐	☐
d) The perimeter of triangle *PQR* is 10.9 cm	☐	☐

The angles in a triangle add up to 180°.

3 *P*, *Q* and *R* are congruent quadrilaterals. Complete shapes *Q* and *R*.

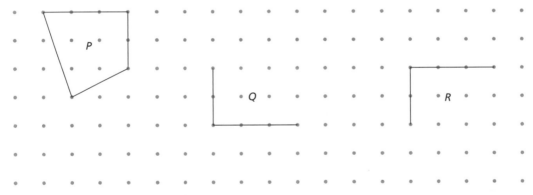

4 Here is a parallelogram *ABCD*. Diagonals can be drawn in *ABCD* to give two triangles. State two pairs of congruent triangles that can be made.

....

5 Paul says these two triangles are congruent.

Explain why he is wrong.

6 cm

Not to scale

6.5 cm

6.5 cm

6 cm

....

....

6 Give one reason why each statement is not true.

a) Two squares are always congruent to each other because they contain the same angles.

....

....

b) Two quadrilaterals with all sides of length 4 cm must be congruent.

....

....

2.2 Circles

Summary of key points

The edge of a circle is called the **circumference**.
A **chord** is a straight line inside a circle joining two points on the circumference.
A **diameter** is a chord passing through the centre of the circle.
A **radius** is a straight line from the centre of a circle to the circumference.
A **tangent** just touches the circumference of a circle without cutting through.

1 **Are these statements true or false?**

	True	False
a) A line that connects two points on the circumference of a circle is called a chord.	☐	☐
b) A line that divides a circle into two equal halves is called a radius.	☐	☐
c) At a point where a tangent to a circle and a diameter meet, the two lines are perpendicular.	☐	☐
d) A diameter is a chord passing through the centre of a circle.	☐	☐

2 **Complete the table to match each circle word to the correct line on the diagram.**

Circle word	Line on diagram
Radius	
Diameter	
Tangent	
Chord	

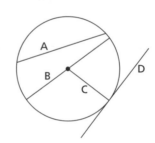

3 **Draw accurate diagrams to match each description:**

a) A circle radius 2.3 cm with a chord 3.5 cm long.

b) A tangent drawn on a circle of diameter 4.2 cm.

c) A circle containing two perpendicular diameters.

a)   b)   c)

4 Give a description of each of the diagrams.

a)

....
....
....
....
....

b)

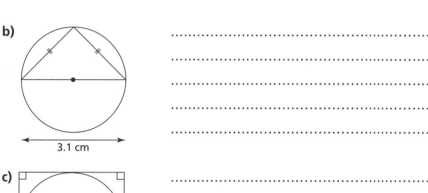

....
....
....
....
....

c)

....
....
....
....
....

5 Andrew drew a circle with radius 4.2 cm. Decide if each of the following statements could be true or cannot be true. Give a reason for each answer.

a) He draws a chord of length 8.7 cm inside his circle.

Could be true ☐   Cannot be true ☐

Reason ....

....

b) He draws a tangent to his circle of length 5 cm.

Could be true ☐   Cannot be true ☐

Reason ....

....

Summary of key points

The diagram shows a square-based pyramid.

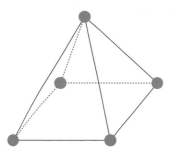

The solid has five **vertices** (shown by the dots).

The solid has five **faces** (one square base and four triangular side faces).

The solid has eight **edges** connecting the vertices.

Exercise 3

1 **Write the name of each shape that is drawn or described.**

a)

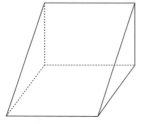

b) A solid with six square faces.

....

c) A solid with two circular faces.

....

....

2 **Complete the table.**

Name of shape	Number of		
	vertices	faces	edges
cuboid	8		
triangular prism		5	
pentagonal pyramid			10

3 **Danny says, 'If a shape has a circular face, then it must have no vertices.'**

Give an example to show that Danny is wrong.

....

4 Draw a line to match each shape to the number of vertices, faces and edges.

10 vertices	8 vertices	4 vertices
7 faces	6 faces	4 faces
15 edges	12 edges	6 edges

5 This solid is formed by placing a pyramid on top of a cube.

This solid has:

.... vertices

.... faces

.... edges

6 Decide if these statements are always true, sometimes true or never true.

	Always	Sometimes	Never
a) The number of edges in a pyramid with a polygon as a base is an even number.	☐	☐	☐
b) The number of faces in a prism is an odd number.	☐	☐	☐
c) The number of faces of a shape is greater than the number of vertices.	☐	☐	☐

Think about

7 Draw a prism with 18 edges. How many vertices and faces does your prism have?

3 Collecting data

You will practice how to:

- Select and trial data collection and sampling methods to investigate predictions for a set of related statistical questions, considering what data to collect (categorical, discrete and continuous data).
- Understand the effect of sample size on data collection and analysis.

3.1 Types of data

Summary of key points

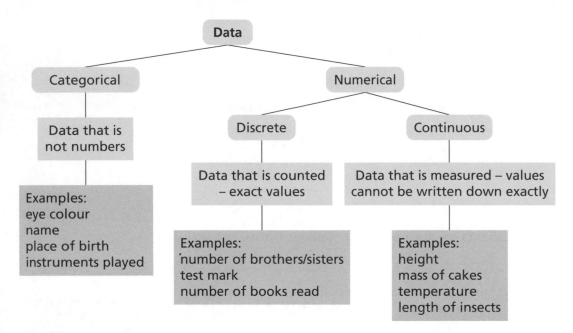

Exercise 1

1 Tick the data sets that contain numerical values.

age	☐	favourite fruit	☐
favourite band	☐	number of cars	☐
volume of water in a bottle	☐		

2 Azil collects information about reading habits. Is each variable categorical, discrete or continuous?

	Categorical	Discrete	Continuous
a) Gender (male/female)	☐	☐	☐
b) Age last birthday	☐	☐	☐
c) Number of books read last week	☐	☐	☐
d) Number of minutes spent reading last week	☐	☐	☐
e) Favourite author	☐	☐	☐

3 Tina collects information about the weather each week in her town. Is each variable discrete or continuous?

a) Amount of rainfall in the week (mm) ....

b) Number of days without rain ....

c) Highest temperature recorded (°C) ....

4 Here are some variables related to houses. Write down whether each variable is discrete or continuous.

a) Number of bedrooms ....　　b) Area of kitchen (m²) ....

c) Length of garden (m) ....　　d) Number of parking spaces ....

Think about

5 Look at some recipes for cakes on the internet. Write down some discrete data and some continuous data relating to each recipe.

3.2 Data collection methods

Summary of key points

To investigate a statistical question, you need to think carefully about:

- what data is needed
- how the data can be collected
- how to organise and present the data.

You often need to collect your own data.

Data could be collected using one of these methods:

- observation
- interviewing people
- using a questionnaire.

Results from observations or interviews can sometimes be recorded on a data collection sheet.

Questions in a questionnaire should be written carefully to avoid bias. They should not try to lead people to giving a particular answer. They should be clear and easy to answer. It is usually easier if the questionnaire includes boxes to tick.

Exercise 2

1 **Obi has been asked to explore the question:**

Do boys play more sport than girls?

He has decided to collect data from people in his class. Which two of these things will it be most useful for him to collect data about?

A	B	C	D
Gender of person	Age of person	Number of hours spent playing sport last week	Favourite sport

.... and ....

2 **Amanda is investigating how much rain different cities have. Which would be the most sensible way for her to collect useful data?**

A	B	C	D
Measure the amount of rainfall each day in her garden	Ask people in her class	Look on the internet	Phone up people in different cities and ask if it is raining

....

3 **Max wants to find out how boys' favourite television programmes compare with girls' favourite television programmes.**

a) Which two of these variables will be most relevant to Max's study?

A	B	C	D
Town where person lives	Gender of person	Number of hours spent watching television last week	Favourite television programme

.... and ....

b) Which of the following would be the most sensible way for Max to collect his data?

A	B	C	D
Ask his brothers and sisters	Ask people in his school	Look up the information in a book	Phone people up randomly

....

4 Angela's teacher has set her this question to investigate.

> Do more cars pass the school each minute in the morning than pass each minute in the afternoon?

What data would Angela need to collect?

....

....

5 Laila wants to ask people which musical instruments they play. Complete the first column to make a suitable data collection sheet.

Instrument	Tally	Frequency
....		
....		
....		
....		
....		
....		

6 A furniture store would like to know what its customers think about the furniture on sale. Suggest a suitable method the shop could use to collect this information.

....

....

....

7 Pam designs a questionnaire about listening to the radio. Here are three of her questions.

Question 1	Question 2	Question 3
Don't you agree that listening to the radio is much better than watching the television?	Which radio station do you listen to the most?	How much time do you spend listening to the radio?

a) What is wrong with the wording of Question 1?

....

....

b) How could Question 2 be improved?

....

....

c) What is wrong with the wording of Question 3?

....

....

8 Hana designs a questionnaire about films. Here is one of her questions.

How many films did you watch last month at the cinema? Tick a box.

1 ☐   2 ☐   3 ☐

a) What is wrong with the tick box choices?

....

....

b) Write better tick boxes for this question.

....

....

9 Ali wants to write a questionnaire to find out what fruit people like best. Write a question that he could have on his questionnaire. Include some tick boxes.

....

....

....

....

....

3.3 Choosing a sample

Summary of key points

Population: The set of all people or things you want to find out about.

Sample: A selection from the population from which data is obtained.

Random sample: A sample is called a random sample if every member of the population has an equal chance of being picked.

The **sample size** is the number of people you decide to survey from the total population. The decision about how much data to collect involves a balance between accuracy and cost. There is greater accuracy if you collect more data, but it is more time consuming to collect.

Exercise 3

1 Phil has been asked to explore the question:

Do girls at this school play more instruments than boys?

a) Which two of these variables should Phil collect data about?

A	B	C	D
Number of instruments played	Age of person	Favourite composer	Gender of person

.... and ....

b) Phil decides to collect data from a sample of 10 children at his school. Comment on his method.

....

....

2 There are 120 people staying in a hotel. The manager wants to choose a sample of these people to ask them their opinion of the hotel.

She selects every 4th person at the hotel.

a) How many people does the manager select? ....

b) Why might the manager prefer to ask a sample of people rather than to ask everyone staying at the hotel?

....

....

3 A company has 300 employees. Fatima wants to choose a sample of 25 of these employees.

a) What is the population for Fatima's investigation?

....

b) Explain how Fatima could choose her sample randomly.

....

....

....

4 200 people attend a concert. Marc wants to ask a sample of these people what they thought about the concert.

He decides to ask 7 people sitting on the back row.

a) Write down two problems with Marc's method.

....

....

b) Suggest a better way for Marc to get the information he wants.

....

....

5 One million people live in a city. Pria wants to ask a random sample of these people what they think about the transport system.

She decides to collect data from one tenth of the people in the city.

Make two comments about Pria's plan.

....

....

Negative numbers and indices

You will practice how to:

- Estimate, add and subtract integers, recognising generalisations.
- Estimate, multiply and divide integers including where one integer is negative.
- Understand the relationship between squares and corresponding square roots, and cubes and corresponding cube roots.

4.1 Adding and subtracting integers

Summary of key points

An **integer** is a whole number.

To add an integer, move along the number line in the **same** direction as the positive or negative sign on the integer.

To subtract an integer, move along the number line in the **opposite** direction from the sign on the integer.

$8 + (–2) = 6$ $8 + (+2) = 10$

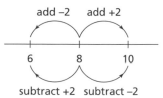

$8 – (+2) = 6$ $8 – (–2) = 10$

Exercise 1

 1–8

1 Complete the table.

Starting temperature	Temperature change	Finishing temperature
3 °C	falls by 7 °C	.........°C
–4 °C	increases by 10 °C	.........°C
.........°C	falls by 12 °C	–8 °C
.........°C	falls by 25 °C	–14 °C
–4 °C	.... by ....°C	–20 °C

Find:

a) $4 - 12 =$ ....

b) $-3 - 10 =$ ....

c) $-14 + 9 =$ ....

d) $-15 + 24 =$ ....

e) $7 - 8 - 3 =$ ....

f) $-11 - 19 + 24 =$ ....

3 **Find:**

a) $3 - (-4) =$ ....

b) $1 + (-7) =$ ....

c) $-9 - (-6) =$ ....

d) $-6 - (+7) =$ ....

e) $-9 + (-3) =$ ....

f) $12 - (-5) + (-4) =$ ....

Draw lines to match calculations with the same answer. There are five matching pairs.

A	B	C	D	E
$3 + (-9)$	$5 - (-4)$	$-12 - (-5)$	$11 + (-2)$	$-3 - 6$

F	G	H	I	J
$8 + (-3)$	$-18 - (-9)$	$-10 + 4$	$-5 - (-10)$	$6 - 13$

5 **In the shape below, all of the columns (vertical lines of numbers) and diagonals have the same sum.**

Fill in the missing numbers.

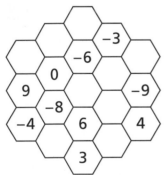

6 **Find the following. You may find it helpful to sketch number lines.**

a) $-67 + 14 =$ ....

b) $-48 - 26 =$ ....

c) $23 - 45 =$ ....

d) $-46 - (-12) =$ ....

e) $-22 - (+26) =$ ....

f) $68 + (-35) =$ ....

7 **The negative integers ■ and ▲ are such that:**

$$■ - 3 = -11 \quad \text{and} \quad ■ - ▲ = -6$$

Find the value of ■ + ▲.

8 *p* is a positive integer and *n* is a negative integer.

Circle the statements that must be true.

A $p + n$ is positive.   B $p - n$ is positive.   C $n - p$ is negative.

9 Use a calculator to find:

a) $208 - (-495)$ b) $-322 + (-713)$ c) $124 - (-476)$

4.2 Multiplying and dividing integers

Summary of key points

The rules for multiplying and dividing integers are:

$+ \times + = +$ $+ \times - = -$ $- \times + = -$

$+ \div + = +$ $+ \div - = -$ $- \div + = -$

Examples: $5 \times (-2) = -10$ $-3 \times 4 = -12$ $-10 \div 5 = -2$ $12 \div (-4) = -3$

Exercise 2

1–7

1 Draw a line to match each calculation with the correct answer.

-3×6		-18
6×3		-2
$-6 \div 3$		2
$6 \div 3$		18

2 Tick the true statements and cross the false statements.

a) $6 \times (-8) = -48$ b) $20 \times (-4) = 5$ c) $-42 \div 7 = -8$

d) $5 \div (-15) = -3$ e) $-10 \times 4 = -40$ f) $6 \times 9 = 54$

3 Find:

a) $-8 \div 4$ .... b) -11×4 ....

c) $27 \div (-3)$ .... d) 9×7 ....

e) $54 \div (-6)$ .... f) -100×7 ....

4 Decide whether each statement is always true, sometimes true or never true.

a) If a and b are integers, then $a \times b$ is greater than a.

....

b) If a and b are positive integers, then $a \times b$ is a positive integer.

....

c) If a is a positive integer and b is a negative integer, then $a \times b$ is a negative integer.

....

d) If a is a positive integer and b is a negative integer, then $b \div a$ is a positive integer.

....

5 Below are two puzzles. In each puzzle, write numbers in the four boxes to make all of the calculations correct.

a)

$\square \times \square = -24$

$\div$ $\div$

$\square \times \square = 6$

$=$ $=$

-2 2

b)

$\square \div \square = -5$

$\times$ $\times$

$\square \times \square = -12$

$=$ $=$

-100 -15

Think about

6 Can you find two possible solutions for question 5 part b)?

7 The sum of two integers is -2 and the product is -24.

Find the two integers.

.... and ....

8 Use a calculator to find:

a) 52 × (−26) ....

b) −98 × 7 ....

c) 1530 ÷ (−45) ....

d) 3198 ÷ 6 ....

e) −1444 ÷ 4 ....

f) −404 × 312 ....

4.3 Estimating

Summary of key points

Estimating the answer to a calculation means making the calculation easier by **rounding** the numbers, and finding an answer that is **approximate** (not exactly correct). In some situations, an approximate answer is good enough.

To estimate the answer to a calculation, round 2-digit numbers to the nearest 10 and 3-digit numbers to the nearest 100 (and 4-digit numbers to the nearest 1000, etc.) Then do the calculation.

You can use the symbol ≈ which means 'approximately equals'.

For example:

$32 × 47 ≈ 30 × 50$ $595 − 812 ≈ 600 − 800$ $−618 ÷ 61 ≈ −600 ÷ 60$

$= 1500$ $= −200$ $= −10$

Exercise 3

1 By rounding each number to the nearest 10, estimate:

a) 61 − 92 ....

b) 27 × (−51) ....

c) −93 ÷ 29 ....

d) −52 + 78 ....

2 By rounding each number to the nearest 100, estimate:

a) 681 − 208 ....

b) −321 × 214 ....

c) 999 ÷ (−99) ....

d) −387 + 613 ....

3 Emma, Faraz and Gregor each estimate 326 × 29. Their working is shown below.

a) Emma's working:

$326 × 29 ≈ 400 × 30$

$= 12\,000$

b) Faraz's working:

$326 × 29 ≈ 300 × 30$

$= 900$

c) Gregor's working:

$326 × 29 ≈ 300 × 20$

$= 6000$

Describe the mistake each person has made.

....

....

....

d) Show how to estimate 326 × 29.

....

4 Use estimation to match each calculation with its correct answer. Draw lines to connect calculations with their answers.

−42 × 36	−3882
3130 − 7012	−5322
−9211 − (−3889)	−6552
91 × (−72)	−1512

5 Three of the calculations below are incorrect. Use estimation to find which calculations are incorrect. Do not calculate the exact answers.

A 96 × (−52) = −492   B −28 × 61 = −1708   C −209 ÷ 19 = −11

D −390 − 504 = −214   E 486 × 41 = 11 226   F 413 − 795 = −382

6 **a)** A school buys 72 calculators costing $11 each. Estimate the total cost of the calculators.

....

b) Company A has 377 employees and Company B has 716 employees. The two companies merge into one new company. Estimate the total number of employees in the new company.

....

c) A website sold 897 books for $22 each. Estimate the total amount paid for the books.

....

7 The estimated answer to a calculation is –400. Write a calculation that could have this estimated answer, using:

a) addition ....

b) subtraction ....

c) multiplication ....

d) division ....

4.4 Indices

Summary of key points

The **square** of a number is the product of two copies of the number.

For example, 4 squared is $4^2 = 4 \times 4 = 16$.

The **cube** of a number is the product of three copies of the number.

For example, 3 cubed is $3^3 = 3 \times 3 \times 3 = 27$.

The **square root** of a number squares to make the number.

For example, the square root of 25 is $\sqrt{25} = 5$ because $5^2 = 25$.

The **cube root** of a number cubes to make the number.

For example, the cube root of 8 is $\sqrt[3]{8} = 2$ because $2^3 = 8$.

The small '2' and '3' that show squaring and cubing are called **indices** or **powers**. The singular of indices is **index**.

Exercise 4   1–9

1 Write the value of:

a) $\sqrt{49}$ **b)** 3^3 **c)** 8^2

....

d) $\sqrt{25}$ **e)** $\sqrt[3]{64}$ **f)** 1^3

....

2 Write the value of:

a) $\sqrt{6} \times \sqrt{6}$ **b)** $\sqrt[3]{(11 \times 11 \times 11)}$ **c)** $\left(\sqrt[3]{99}\right)^3$

....

3 Write whether each statement is true or false.

a) $21^2 > 20^2$

....

b) $\sqrt{22} < 5$

....

c) $6^3 > 125$

....

d) $\sqrt[3]{30} < 3$

....

e) $10.1^2 < 100$

....

f) $42^2 < 42^3$

....

4 Amana and James both think of a positive number.

The square root of Amana's number is 9. The square of James's number is 9.

Find the sum of Amana's number and James's number.

....

5 Jake works out the values of some square numbers and cube numbers. His results are shown below.

$2^2 = 4$

$3^2 = 6$

$5^3 = 15$

Describe the mistake Jake has made.

....

Think about

6 Jake made the same mistake in all three calculations. Why did he get the correct answer for one of them?

7 Find two square numbers that add together to make:

a) 20 .... and ....

b) 130 .... and ....

8 Jenna thinks of a square number less than 100.

Doug thinks of a cube number less than 100.

The difference between their numbers is 15.

Work out which number they each thought of.

Jenna's number .... Doug's number ....

9 Joe cannot remember the door number of his friend's house. He remembers that it is a cube number, and that it is greater than 10 but less than 70. It is also a multiple of 9. What could the door number be?

....

10 Use a calculator to find:

a) $\sqrt{225}$

....

b) 21^3

....

c) 101^2

....

d) $\sqrt{4096}$

....

e) $\sqrt[3]{1728}$

....

f) 7^3

....

5 Expressions

You will practice how to:

- Understand that letters can be used to represent unknown numbers, variables or constants.
- Understand that a situation can be represented either in words or as an algebraic expression, and move between the two representations (linear with integer coefficients).
- Understand that the laws of arithmetic and order of operations apply to algebraic terms and expressions (four operations).

5.1 Letters for numbers

Summary of key points

In algebra, a letter is used to represent an unknown number or measurement (a **variable**).

An **expression** is a mathematical statement about a variable. For example, if you have a number, t, then add 5, you could write $t + 5$.

In $t + 5$, t and 5 are both **terms** in the expression. t is a variable and 5 is a constant term in the expression.

The expression $5f$ represents $5 \times f$. The 5 is called the **coefficient** of f.

An **equation** is a mathematical statement where an expression is made equal to something. For example, $t + 5 = 12$. Equations can be solved to find the value of the unknown variable.

Exercise 1

1 **Rewrite the number puzzles as equations by using a letter to represent the missing number.**

	Number puzzle	Equation
Example	26 + ... = 100	26 + t = 100
a)	11 + ... = 20	....
b)	.... − 2 = 15	....
c)	.... × 10 = 240	....
d)	45 = ... × 9	....

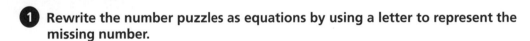

2 Draw lines to match each statement to the correct algebraic expression.

'I think of a number and add 7.'

'I think of a number and multiply it by 7.'

'I think of a number and subtract 7.'

'I think of a number and divide it by 7.'

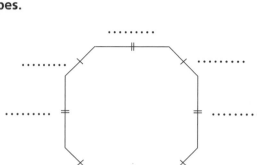

3 Use letters to represent the side lengths of these shapes.

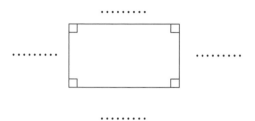

4 Here are some statements.

$c + 3 = 11$
This is known as an equation.

$f + g + 2$
This involves two terms, f and g.

A parcel has mass $(m + 2)$ kg.
This is an expression.

$4c = 12$
The coefficient of c is 12

Tick the statements that are true.

For the statements that are false, write the true statement.

Think about

5 Give examples to explain each of these algebraic words: variable, equation, expression and term.

5.2 Forming expressions

Summary of key points

Example

A shop sells these clothes.

The price of the scarf is c dollars.

The jumper costs twice as much as the scarf.

The price of the jumper can be written as 2c dollars.

The trainers cost 10 dollars more than the jumper.

An expression for the cost of the trainers is 2c + 10 dollars.

Exercise 2

1 A shop sells boxes of crayons. Each box contains *b* crayons.

Write down the number of crayons in each picture.

b crayons

a)

.... crayons

b)

.... crayons

c)

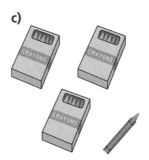

.... crayons

2 An apple costs *a* cents. A banana costs *b* cents. Write expressions for the cost of the fruit in each picture.

a)

.... cents

b)

.... cents

c)

.... cents

d)

.... cents

3 Nathan thinks of a number. Call his number *n*. Write an expression in terms of *n* for the result when you:

a) add 2 to Nathan's number ....

b) subtract 4 from Nathan's number ....

c) multiply Nathan's number by 9 ....

d) multiply Nathan's number by 4 and then add 1 ....

4 A pen costs *p* dollars. Write an expression in terms of *p* for the cost of 6 pens.

.... dollars

5 A hat costs *h* dollars. It is reduced by 4 dollars. Write an expression in terms of *h* for the new cost of the hat.

.... dollars

6 There are *c* cakes in a packet. A boy eats 3 cakes. Write an expression in terms of *c* for how many cakes are left.

.... cakes

7 Sheena has *n* badges. She shares them equally between her two children.

Write an expression in terms of *n* for how many badges each child has.

.... badges

8 The cost of train tickets is:

Child ticket $*c*   Adult ticket $*a*

What is the total cost of 5 child tickets and 2 adult tickets? $..............

9 Four people are talking about how old they are. Clara's age is *c* years. Write an expression in terms of *c* for the age of all the other people.

Clara		*c* years
Dimitri	'I am two years older than Clara.'	.... years
Eva	'I am three times as old as Clara.'	.... years
Faiyaz	'I am six years younger than Eva.'	.... years

10 Here is a triangle with side lengths as marked on the diagram.

Write an expression for the perimeter of the triangle.

t cm   *u* cm   4 cm

> The perimeter is the total distance around the edge of a shape.

.... cm

11 A shop sells *b* loaves of bread on Monday. Complete the table to show information about bread sales on other days.

Day	Description of bread sales	Expression for bread sales
Tuesday	Ten more loaves sold than on Monday	....
Wednesday	.... loaves sold than on Monday	$b - 5$
Thursday	Twice as many loaves sold as on Monday	....
Friday	.... more loaves sold than on	$2b + 20$

12 *n* is a positive whole number.

Put a ring around the numbers that are **greater** than *n*.

$n + 2$ $n - 3$ $6n$ $n \div 2$ $n + 0$

5.3 Order of algebraic operations and substitution

Summary of key points

In arithmetic there are certain rules that are always true, for example:

$7 + 2 = 2 + 7$

These rules of arithmetic are also true in algebra.

When you want to find an expression for a particular value, you need to **substitute** that value into the expression, then use the correct order of **operations**.

For example, if $m = 5$

• $11m = 11 \times 5 = 55$ • $30 - m = 30 - 5 = 25$

Exercise 3

1 Are these statements true or false?

	True	False
a) $1 - 5 = 5 - 1$	☐	☐
b) $4 \times 8 = 8 \times 4$	☐	☐
c) $10 \div 2 = 2 \div 10$	☐	☐
d) $19 + 82 + 3 = 82 + 3 + 19$	☐	☐
e) $5 + 8 - 2 = 8 - 2 + 5$	☐	☐

2 Circle the statements that are true.

$f \times g = g \times f$ $t \div 2 = 2 \div t$ $20 - x = x - 20$ $x + y - 5 = 5 - x - y$

$a + b - c = a - c + b$ $v \div u = u \div v$ $m + 2 = 2 + m$ $p \times q \times p = q \times p \times p$

3 Complete these statements to make them true.

a) $a + b + c = b + \text{........} + c$

b) $2 \times m \times n = \text{........} \times m \times 2$

c) $6 + c - d = c - d + \text{........}$

d) $t \times u \times v = v \times t \times \text{........}$

4 Form three matching pairs from the expressions in the box.

$a \times b \times 3$	$3 \times b \times a$	$b - 3 + a$	$b + 3 + a$
$3 + a - b$	$a \times 3 \times a$	$a + b + 3$	$b + a - 3$

.... = ... = ...

.... = ...

5 Find the value of each expression if $m = 6$.

a) $m + 4$ ....

b) $2m$ ....

c) $m - 2$ ....

d) $3m + 1$ ....

e) $10 - m$ ....

f) $5m - 7$ ....

6 Find the value of each expression if $t = 12$ and $u = 4$.

a) $u + 3$ ....

b) $t - u$ ....

c) $20 - t$ ....

d) $\frac{t}{4} - 7$ ....

e) $5u$ ....

f) $2t + u$ ....

g) $\frac{u}{2}$ ....

h) $3u - t$ ....

i) $\frac{t}{3}$ ....

j) $2t + 3u$ ....

7 If $p = 20$ and $q = 7$, find the value of:

a) $2q - 6$ ....

b) $5p - 11$ ....

c) $4q - 9$ ....

d) $\frac{p}{4} - 8$ ....

8 $x = 6$ $y = 10$

Circle the expression which gives a different value to the other two.

a) $y + 16$ $2x$ $3y - 4$

b) $\frac{x}{2}$ $y - 7$ $\frac{y}{5}$

c) $x + y$ $3x$ $2y - 2$

d) $3y - 4x$ $5x - 2y - 4$ $3x + y - 12$

9 If $z = 4$, write in the missing number to make each statement correct.

a) $3z + $ .... $= 17$ **b)** $5z - $ .... $= 18$ **c)** .... $- z = 15$

10 Carla designs a spreadsheet.

	A	B	C
1	8	= A1*4	
2	11	= A2 + 3	
3	3	= A3 − 2	= B1 + B2 + B3

Find the value that will be shown in cell C3.

Think about

11 If $n = 8$, write four different expressions that have a value equal to 20.

Example: $3n - 4 = 20$

6 Symmetry

You will practice how to:

- Identify, describe and sketch regular polygons, including reference to sides, angles and symmetrical properties.
- Identify reflective symmetry and order of rotational symmetry of 2D shapes and patterns.

6.1 Polygons

Summary of key points

A **regular polygon** has these properties:

- It is a closed shape made of straight sides.
- All the sides are the same length.
- All the angles are the same size.

The number of sides a regular polygon has tells you the number of lines of symmetry and the order of rotational symmetry.

Example: A **square** has 4 lines of symmetry and rotational symmetry of order 4.

Exercise 1

1 **Kieron is describing a shape.**

'It has 5 sides, all equal in length. All the angles are the same.'

Write the name of the shape Kieron is describing.   ....

2 **Alia is describing a shape.**

'It has 8 sides. There are 8 lines of symmetry.'

Write the name of the shape Alia is describing.   ....

3 **Complete the table to show the number of lines of symmetry and the order of rotational symmetry of each regular polygon.**

Regular polygon	Number of lines of symmetry	Order of rotational symmetry
Hexagon		
Heptagon		
Nonagon		
Decagon		

4 A dodecagon is a polygon with twelve sides. Describe the symmetries you would expect a regular dodecagon to have.

....

....

5 The diagram shows how a regular heptagon can be divided to make two pentagons by drawing one line across it.

Show how a regular heptagon can be divided to make an isosceles triangle and an octagon.

6 Show how a regular octagon can be divided into two isosceles triangles and a decagon.

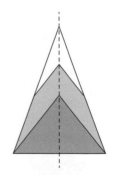

6.2 Line and rotation symmetry

Summary of key points

There are two types of symmetry: **line** symmetry and **rotational** symmetry.

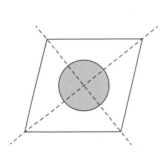
2 lines of symmetry
Rotational symmetry order 2

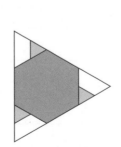
No lines of symmetry
Rotational symmetry order 3

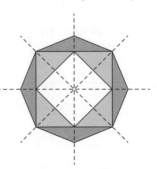
4 lines of symmetry
Rotational symmetry order 4

1 line of symmetry
No rotational symmetry

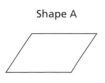

1 Here are eight shapes.

Shape A   Shape B   Shape C   Shape D

Shape E   Shape F Shape G 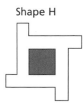 Shape H

Complete the table to show the order of rotational symmetry for each shape.
One shape has been done for you.

Order of rotational symmetry			
Order 1	**Order 2**	**Order 3**	**Order 4**
	A		

2 Describe the line symmetry and rotational symmetry of each shape.

a)

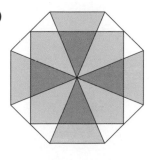

....

b)

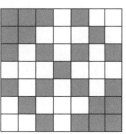

....

c)

....

d)

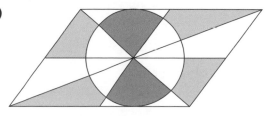

....

3 Here are six shapes.

A

B

C

D

E

F

Complete the table to show how many lines of symmetry and the order of rotational symmetry for each shape.

	A	B	C	D	E	F
Number of lines of symmetry						
Order of rotational symmetry						

4 Colour in eight squares on each shape to make a shape with the given number of lines of symmetry.

a)

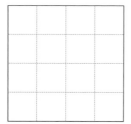

(4 lines of symmetry)

b)

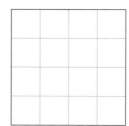

(2 lines of symmetry)

c)

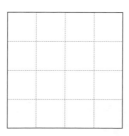

(1 line of symmetry)

d)

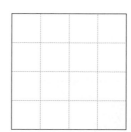

(0 lines of symmetry)

5 Colour in nine triangles on each shape to make a shape with the number of lines of symmetry shown.

a)

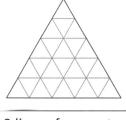

(3 lines of symmetry)

b)

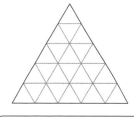

(1 line of symmetry)

c)

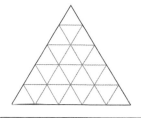

(0 lines of symmetry)

6 Shade six triangles so that the shape has:

a) 3 lines of symmetry and rotational symmetry of order 3

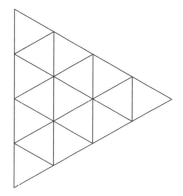

b) no line symmetry and rotational symmetry of order 3

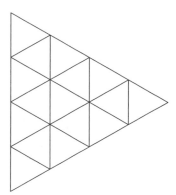

Think about

7 Draw four copies of this shape.

Shade two sections in each drawing and write down all the symmetries of each shape.

Try to shade your shapes so that the symmetries in each shape are unique.

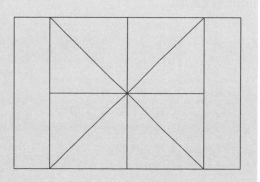

7 Rounding and decimals

You will practice how to:

- Use knowledge of place value to multiply and divide whole numbers and decimals by any positive power of 10.
- Round numbers to a given number of decimal places.
- Estimate, add and subtract positive and negative numbers with the same or different number of decimal places.
- Estimate, multiply and divide decimals by whole numbers.

7.1 Multiplying and dividing by positive powers of 10

Summary of key points

10^6 is 'the sixth power of 10' or '10 to the power of 6'. The 6 is the **index** or **power** of 10.

A positive integer power of 10 tells you how many copies of 10 to multiply together. For example, $10^6 = 10 \times 10 \times 10 \times 10 \times 10 \times 10$.

When the index is zero, $10^0 = 1$. When the index is 1, $10^1 = 10$.

Multiplying by powers of 10	Dividing by powers of 10
To multiply by a power of 10, move each digit to the left the number of places shown by the index.	To divide by a power of 10, move each digit to the right the number of places shown by the index.
For example:	For example:

$1.053 \times 10^2 = 105.3$

100	10	1	•	0.1	0.01	0.001
		1	•	0	5	3
1	0	5	•	3		

$6.1 \div 10^3 = 0.0061$

1	•	0.1	0.01	0.001	0.0001
6	•	1			
0	•	0	0	6	1

Exercise 1

1–5 and 7–9

1 Jack has ten number cards.

A	B	C	D	E
10^2	one million	$10 \times 10 \times 10$	10^6	one hundred thousand

F	G	H	I	J
10^5	10^1	10^3	100	ten

Find the pairs of cards with the same value.

……….. and ………..   ……….. and ………..   ……….. and ………..

……….. and ………..   ……….. and ………..

2 **a)** A piece of paper is 0.06 mm thick. There are 1000 sheets of paper in a pile.

Find the height of the pile.   ……………………… mm

> How would you convert your answer to part **a)** into centimetres, or metres?

b) Takumi has 100 identical coins. The total mass of all the coins is 1385 g.

Find the mass of each coin.   ……………………… g

3 **One billion is one thousand million. Write one billion as a power of 10.** ……………

4 **Write the missing number in each calculation.**

a) $64.2 \times 10^{\cdots} = 642$ **b)** $0.025 \times 10^3 =$ ………………

c) ……………… $\times 10^2 = 650$ **d)** $10^3 \times$ ……………… $= 945$

e) $56 \div 10^4 =$ ………………   **f)** $341 \div 10^{\cdots} = 0.0341$

g) ……………… $\div 10^3 = 4.63$ **h)** ……………… $\div 10^2 = 8.47$

i) $45 \times 10^4 =$ ……………… $\times 10^3$ **j)** $0.062 \times 10^{\cdots} = 620 \div 10^3$

5 **Find:**

a) 42×10^4 ………………   **b)** 0.851×10^5 ………………

c) 1.402×10^2 ………………   **d)** $865 \div 10^2$ ………………

e) $76.25 \div 10^3$ ………………   **f)** $4 \div 10^5$ ………………

6 **Use your calculator to find:**

a) $365 \div 10^3$ ………………   **b)** 0.165×10^6 ………………

c) 0.0001×10^3 ………………   **d)** $26.81 \div 10^2$ ………………

7 **Orla says, 'To multiply a number by 10^3, write three 0s on the end of the number.'**

a) Give an example to show that Orla's rule sometimes works.

………………………………………………………………………………………………………

b) Give an example to show that Orla's rule does not always work.

………………………………………………………………………………………………………

c) Describe when you can and cannot use Orla's rule.

....

....

 8 *m* is a number greater than zero. Write the following calculations in order of the size of their answer. Write the smallest first.

a) $m \times 10^3$ b) $m \div 10^3$ c) $m \times 10^4$ d) $m \div 10^4$

....

Think about

9 Write six different calculations with the answer 72.5. Each calculation should involve multiplying or dividing a number by a power of 10.

7.2 Rounding to decimal places

Summary of key points

You can round a number to a specified number of **decimal places** (d.p.). The table below shows examples.

Number	Rounded to 1 d.p.	Rounded to 2 d.p.	Rounded to 3 d.p.	Rounded to 4 d.p.
3.60187	3.6	3.60	3.602	3.6019
2.153	2.2	2.15	2.153 (no rounding needed)	not possible
62.99999	63.0	63.00	63.000	63.0000

Rounding to more places makes the rounded number more **accurate**.

Exercise 2

1, 3, 4 and 6–8

1 Complete the table by rounding the numbers to the accuracy stated.

		Rounded to 1 d.p.	Rounded to 2 d.p.	Rounded to 3 d.p.
a)	2.81594			
b)	0.06579			
c)	7.83057			
d)	68.89364			

Use your calculator to do these calculations. Round each answer to three decimal places.

a) $4 \div 6$ ....

b) $3 \div 13$ ....

c) $9 \div 16$ ....

d) $22 \div 51$ ....

e) $\sqrt{8}$ ....

f) 12.59^2 ....

g) $\sqrt{20}$ ....

h) $\sqrt[3]{10}$ ....

3 Jana is asked to write 0.22354 correct to 3 d.p. Her answer is 0.223. Is her answer correct? Explain your answer.

....

....

4 Takahiro rounds 0.8001 to three decimal places. His answer is 0.8. Explain why his answer is wrong, and write the correct answer.

....

....

5 Use a calculator to solve these problems.

a) The length of the side of a square is 62.71 cm. Calculate the area of the square, rounded to one decimal place.

.... cm^2

b) A piece of rope of length 12.70 m is cut into 8 equal pieces. Calculate the length of each piece, rounded to two decimal places.

.... m

6 Alma is rounding the number 46.41319 to three decimal places. She says, 'Because the last digit is a 9, I need to round up the 3 to a 4. The answer is 46.414.'

Do you agree with Alma? Explain your answer.

....

....

7 Amir thinks of a number that has two decimal places. When he rounds the number to one decimal place, it is 2.5. When he rounds the number to the nearest integer, it is 2.

Work out a possible value for Amir's number.   ....

8 A runner completed a 100 m race in a time of 11.39 seconds.

a) Give an example of a time with three decimal places that rounds to 11.39 s to two decimal places.

.... s

b) How many different times with exactly three decimal places round to 11.39 s? (You do not have to list them all, but you may find it helpful.)

....

....

7.3 Adding and subtracting decimals

Summary of key points

The table below shows how to use column addition and subtraction if there is a negative number in the question or the answer.

Type of calculation	Subtracting a bigger positive number from a smaller positive number	Adding a smaller negative number to a bigger positive number	Adding a bigger negative number to a smaller positive number	Adding two negative numbers, or subtracting a positive number from a negative number
Examples	$2.35 - (+5.4)$ $2.35 - 5.4$	$-2.35 + 5.4$ $5.4 + (-2.35)$	$-5.4 + 2.35$ $2.35 + (-5.4)$	$-2.35 + (-5.4)$ $-2.35 - 5.4$ $-2.35 - (+5.4)$
Sign of answer	negative	positive	negative	negative
Written method	Calculate $5.4 - 2.35$, which is the negative of $2.35 - 5.4$ $$\begin{array}{r} 3\,_1 \\ 5.\cancel{4}0 \\ -\ 2.35 \\ \hline 3.05 \end{array}$$ $2.35 - 5.4 = -3.05$	Rewrite as a subtraction: $5.4 - 2.35$ $$\begin{array}{r} 3\,_1 \\ 5.\cancel{4}0 \\ -\ 2.35 \\ \hline 3.05 \end{array}$$ $-2.35 + 5.4 = 3.05$	Rewrite as $2.35 - 5.4$. Calculate $5.4 - 2.35$, which is the negative of $2.35 - 5.4$ $$\begin{array}{r} 3\,_1 \\ 5.\cancel{4}0 \\ -\ 2.35 \\ \hline 3.05 \end{array}$$ $2.35 - 5.4 = -3.05$	Calculate $2.35 + 5.4$, which is the negative of $-2.35 - 5.4$ $$\begin{array}{r} 2.35 \\ +\ 5.40 \\ \hline 7.75 \end{array}$$ $-2.35 - 5.4 = -7.75$

1 Find, either in your head or by sketching a number line:

a) 0.6 – 0.9

b) –4.7 – 7.4

....

....

c) 2.9 – (+3.1)

d) –3.2 + 5.6

....

....

2 Fill in the table. Show what column addition or column subtraction to do for each calculation, and state whether to write a negative sign on the result.

Calculation	Column addition or subtraction to do	Write a negative sign on the result?
a) –3.86 + 2.5	3.86 – 2.5	Yes
b) –12.3 + 15.1		No
c) 0.87 + 0.226	0.87 + 0.226	
d) –6.59 – 4.26		
e) 0.67 + –0.5		
f) 2.08 – 3.75		

3 For each calculation below, estimate the answer. Then use a written method to calculate the exact answer.

a) –11.21 – 8.6

b) 1.25 – 5.73

....

....

c) –45.431 + 12.28

d) –9.28 – (–7.108)

....

....

4 Here are some number cards:

-0.91 0.63 -0.88 -0.54 -0.63

a) Using each card no more than once, fill in the calculation

☐ + ☐ to make

  i) the highest possible answer ....

  ii) the lowest possible answer ....

b) Using each card no more than once, fill in the calculation

☐ − ☐ to make

  i) the highest possible answer ....

  ii) the lowest possible answer ....

5 Find the missing numbers.

a) $-5.6 + \text{...................} = -0.9$

b) $-0.28 + \text{...................} = -0.42$

c) $\text{...................} - (-9.845) = 5.488$

d) $3.56 - 2.81 - \text{...................} = -3.84$

6 Carey wants to buy a jacket for $22.95, a shirt for $16.78 and a jumper for $25.99.

She has $58.55, which is not enough to buy all three items.

Find how much more money she needs.

$.....................

7 In the calculation $\bullet - \blacksquare = -0.8$, the missing numbers each have one decimal place. Write possible values of the missing numbers if:

a) they are both negative ....

b) they are both positive ....

c) one number is negative and the other number is positive ....

8 Use each of the digits 1, 2, 4, 7 and 8 once each to make this statement true.

$$0.\square\square - 0.\square\square\square = 0.569$$

Think about

9 Make up your own word problems that can be solved by adding and/or subtracting decimals.

Summary of key points

To multiply an integer by a decimal using long multiplication, first ignore the decimal point. For example, to find 2.819 × 56, calculate 2819 × 56. The result is 157 864. To place the decimal point in the answer, use one of the three methods below.

| Estimate the answer:

2.819 × 56 ≈ 3 × 60 = 180

The exact answer is 157.864 (because this is nearest to 180). | Count the digits after a decimal point in the question. There are three, in 2.**819**.

Place the decimal point three digits to the left of the last digit of the result: 157.**864** | Notice that 2.819 is 1000 times less than 2819.

2819 × 56 = 157 864

↓ ÷ 1000   ↓ ÷ 1000

2.819 × 56 = 157.864 |

When you divide a decimal by an integer, write a decimal point in the answer above the decimal point in the divisor. You can use **short division** (working out remainders in your head) or **long division** (working out remainders on paper).

Exercise 4

1. **Write whether each statement is true or false.**

   a) 0.3 × 4 = 0.12

   b) −0.06 × 5 = −0.30

   c) 0.25 × 6 = 1.5

   d) 0.42 ÷ 7 = 0.6

   e) 0.02 ÷ −4 = 0.005

   f) 0.333 ÷ 11 = 0.03

2. **Write the missing numbers.**

   a) 0.28 ÷ 7 = ...

   b) 6 × 0.04 = ...

   c) 0.64 ÷ ....= −0.08

   d) 0.25 × .... = 1

   e) 15 × .... = −0.15

   f) .... ÷ 4 = 0.008

3. **For each calculation, estimate the answer by rounding each number.**

   Then calculate the exact answer using a written method.

   a) 0.122 × 56

   b) 5.281 × 17

   ....

   ....

   c) −8.56 ÷ 4

   d) 3.335 ÷ 23

   ....

   ....

4 Marcel wants to find 3.59 × 874.

He uses a written method to calculate 359 × 874 = 313 766.

a) Write the answer to 3.59 × 874 by adding a decimal point in the correct place.

....

b) Explain how you know where to write the decimal point.

....

5 For each calculation, estimate the answer by rounding each number.

Then calculate the answer using a written method. Round each answer to three decimal places.

a) 58.6 ÷ 6   b) 7.513 ÷ 8

....

c) 85.29 ÷ 23   d) 71.841 ÷ −61

....

6 Ben has one roll of string that is 1.44 m long. He cuts it into two equal parts. Then he cuts the first half into 18 equal pieces. He cuts the second half into 12 equal pieces.

Which pieces are longer? How much longer are they?

....

7 If 2.367 ÷ ● = ■, and ● is a single-digit number, what are the lowest and highest possible values of ■?

....

8 a) Calculate 5.387 × 6 using a written method.

....

> Use your times table knowledge and divisibility tests.

b) Without doing any calculations, write down three related multiplication facts.

....

....

c) Without doing any calculations, write down three related division facts.

....

....

8 Presenting and interpreting data 1

You will practice how to:

- Record, organise and represent categorical, discrete and continuous data. Choose and explain which representation to use in a given situation:
  - o Venn and Carroll diagrams
  - o tally charts, frequency tables and two-way tables
  - o dual and compound bar charts
  - o frequency diagrams for continuous data
- Interpret data, identifying patterns within and between data sets, to answer statistical questions. Discuss conclusions, considering the sources of variation, including sampling, and check predictions.

8.1 Two-way tables

Summary of key points

Two-way tables allow you to organise data for two different variables using columns and rows.

Exercise 1

1 Emilie is investigating the number of pieces of fruit that boys and girls ate on Monday. The two-way table shows her results.

Number of pieces of fruit	0	1	2	3	4	5	More than 5
Boys	7	4	8	5	2	1	1
Girls	4	9	6	8	0	3	2

a) How many girls did Emilie collect data from? ....

b) How many boys ate exactly two pieces of fruit on Monday? ....

c) What is the total number of boys and girls that ate no fruit on Monday? ....

d) How many more girls than boys ate at least three pieces of fruit on Monday? ....

2 Phil records the number of cars in a car park on 26 Mondays and 26 Tuesdays.

Number of cars	21–25	26–30	31–35	36–40	41–45
Monday	3	2	1	3	17
Tuesday	4	3	2	6	11

a) On how many Mondays were there at least 36 cars in the car park? ....

b) Phil says that the car park had less than 36 cars parked on over half of the Tuesdays. Is he correct? Explain your answer.

....

....

3 Madge sells socks in her store. The table shows the colours of socks bought by male and female customers.

	Red	White	Black	Blue	Grey	Total
Male	7	6	15			40
Female			4		6	
Total		14		10	13	75

Complete the table.

4 A headteacher is investigating the link between pupil absence and Science test scores. She collects data for 100 children in this two-way table.

		Mark			
		21–25	26–30	31–35	36–40
	0–2	2	6	9	7
Number of days absent	3–5	5	7	15	7
	6–8	6	11	12	3
	9 or more	4	3	3	0

a) How many children who scored a mark in the range 31–35 were absent for no more than 2 days? ....

b) How many pupils in total were absent for 9 or more days? ....

c) How many pupils in total scored 21–25 marks? ....

d) The headteacher says over half of the children who were absent for six or more days scored 30 marks or less in the test. Is the headteacher correct? Give reasons for your answer.

....

....

....

....

5 Vanessa makes small and large cakes. She decorates her cakes with either chocolates or sweets.

The Venn diagram shows information about the cakes she made yesterday.

a) How many cakes did Vanessa make in total yesterday? ....

b) Find the number of cakes she decorated with chocolates. ....

c) Complete the table to show the information.

	Small	Large	Total
Sweets			
Chocolate			
Total			

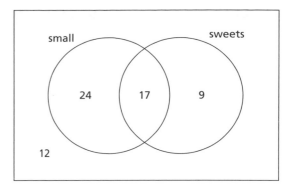

8.2 Dual and compound bar charts

Summary of key points

Dual bar chart: A bar chart for showing data from two groups side by side.

Compound bar chart: A bar chart showing data for two or more groups with the bars stacked on top of each other.

Exercise 2

1 The dual bar chart shows some information about the tickets sold for a concert.

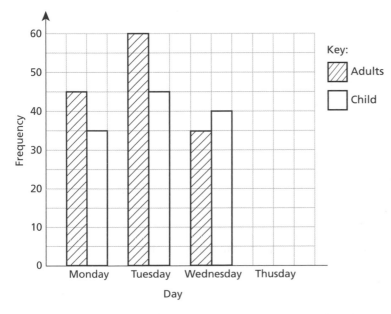

Key:
▨ Adults
☐ Child

a) Write down how many adult tickets were sold for the concert on Wednesday.

....

b) Find the total number of tickets that were sold for the concert on Monday.

....

c) Compare the number of child tickets sold on Tuesday with the number of child tickets sold on Monday.

....

....

d) Here are the ticket sales for Thursday's concert:

55 adult tickets   25 child tickets

Complete the dual bar chart.

2 A school organises an activity day. The children can choose to do either circus skills or water sports. The table shows the number of males and females choosing each activity.

	Circus skills	Water sports
Males	6	11
Females	13	7

a) Draw a compound bar chart to show the information. Remember to include a key.

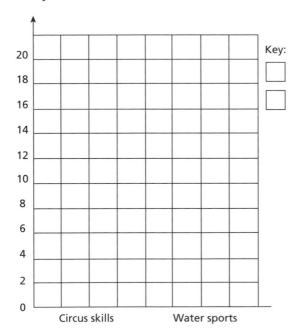

b) Give a reason why a compound bar chart is a suitable diagram for the information.

....

....

3 A hotel has bedrooms on the ground floor and the first floor. The rooms are either standard or luxury. The compound bar chart shows the numbers of each type of bedroom in the hotel.

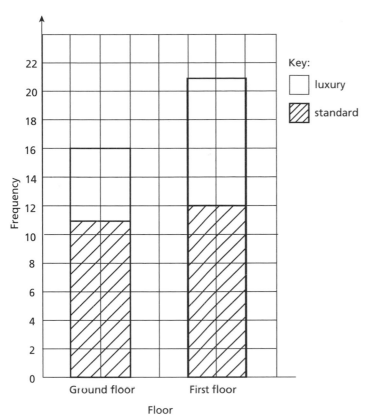

Tick to show if each statement is true or false.

	True	False
a) There are 16 bedrooms on the ground floor.	☐	☐
b) There are 12 luxury bedrooms on the first floor.	☐	☐
c) There are more bedrooms on the ground floor than on the first floor.	☐	☐
d) The total number of standard bedrooms in the hotel is 23.	☐	☐

4 Lara knits scarves and jumpers. The dual bar chart shows the number of each she knitted in 2018.

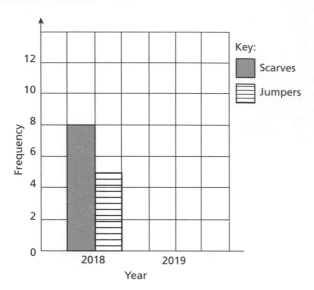

In 2019,

- Lara knitted the same number of items as she did in 2018
- she knitted fewer jumpers than she did in 2018.

Complete the dual bar chart by drawing a possible pair of bars for 2019.

8.3 Frequency diagrams for continuous data

Summary of key points

The frequency diagram shows the length (in metres) of a random sample of lions.

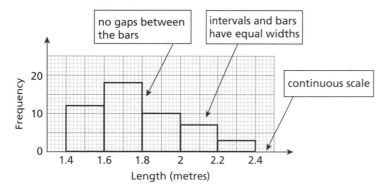

- The total number of lions in the sample is $12 + 18 + 10 + 7 + 3 = 50$
- The fraction of lions in the sample that have length greater than 2 m is $\frac{10}{50} = \frac{1}{5}$.
- So about $\frac{1}{5}$ of lions in the whole population should be greater than 2 m in length.

1 The frequency diagram shows the time that some students took to complete their homework.

The bar for **2–2.5** hours is missing.

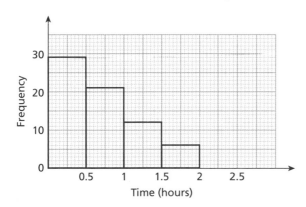

a) Find how many students took less than 30 minutes to complete their homework.

....

b) 9 students took **2–2.5** hours to complete their homework. Complete the frequency diagram.

c) Find the total number of students.

....

2 Sloane records the temperature (in °C) outside his house on 46 different days.

18.2	19.4	16.0	17.5	20.7	16.3	18.3	18.4	19.2	18.4
20.6	16.7	18.3	19.6	17.1	18.1	17.4	16.5	18.9	18.7
16.9	17.7	18.2	18.6	19.7	17.6	16.5	17.8	19.4	17.8
20.3	19.3	19.5	18.9	17.1	17.6	18.5	17.8	18.7	19.3
16.2	18.5	18.9	19.4	19.7	20.6				

a) Complete the frequency table for Sloane's data.

Temperature (°C)	Tally	Frequency
16–17		
17–18		
18–19		
19–20		
20–21		

b) Draw a frequency diagram to illustrate Sloane's data.

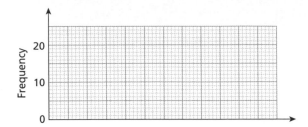

3 8000 people are watching an athletics competition in a stadium.

Niki takes a random sample of 100 of these people.

She asks them how far they travelled to get to the stadium.

Her results are shown in the frequency diagram.

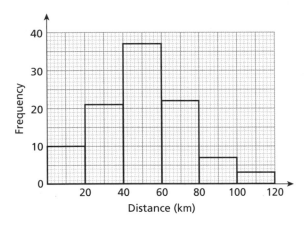

a) How many people in the sample travelled between 40 km and 60 km?

....

b) What fraction of the people in the sample travelled more than 60 km?

....

c) Calculate an estimate of the number of people in the whole stadium who travelled less than 20 km.

....

Think about

4 There are 60 members of a club.

- One quarter of the members are younger than 25 years old
- One fifth of members are aged 75–100 years
- More members are aged 25–50 years than are aged 50–75 years.

Draw a possible frequency diagram to show the ages of all members of the club.

9 Fractions

You will practice how to:

- Estimate and add mixed numbers, and write the answer as a mixed number in its simplest form.
- Estimate, multiply and divide proper fractions.

....

9.1 Adding mixed numbers

Summary of key points

To add mixed numbers:

1) Add the fractional parts, changing to a common denominator if necessary.

2) Add on the whole numbers.

Example: Find $2\frac{2}{3} + 1\frac{3}{4}$.

$$\frac{2}{3} + \frac{3}{4} = \frac{8}{12} + \frac{9}{12} = \frac{17}{12} \text{ or } 1\frac{5}{12}$$

$$1\frac{5}{12} + 2 + 1 = 4\frac{5}{12}$$

To estimate the sum of mixed numbers, you can round each number to the nearest integer and then add.

Example: Estimate $1\frac{2}{5} + 3\frac{2}{3}$.

$$1\frac{2}{5} + 3\frac{2}{3} \approx 1 + 4 = 5$$

Exercise 1

1) **Estimate the answers to these additions, then calculate the exact answers. Give each answer as a mixed number in its simplest form.**

a) $1\frac{2}{3} + \frac{1}{12} =$ b) $2\frac{3}{4} + \frac{2}{5} =$ c) $1\frac{5}{6} + 1\frac{4}{9} =$

....

d) $3\frac{3}{5} + 2\frac{1}{8} =$ **e)** $2\frac{5}{6} + 4\frac{3}{8} =$ **f)** $5\frac{7}{10} + 2\frac{3}{4} =$

....

2 A baby snake is $15\frac{3}{4}$ cm long. It grows another $5\frac{7}{10}$ cm. Calculate its new length.

....

....

3 Estimate the answers to these additions, then calculate the exact answers. Give each answer as a mixed number in its simplest form.

a) $1\frac{1}{2} + 2\frac{3}{8} + 1\frac{3}{4} =$ **b)** $2\frac{1}{3} + 1\frac{7}{12} + \frac{3}{8} =$

....

4 Use the digits 1, 2, 3, 4, 5 and 6 once each to complete this calculation.

$$\square\frac{\square}{\square} + \square\frac{\square}{\square} = 8\frac{19}{20}$$

Think about

5 Write some sentences explaining to a friend how to add mixed numbers. Include some examples to support your explanations.

Summary of key points

Multiplying proper fractions

You can simplify either before or after multiplying. For example:

Simplifying after multiplying: Simplifying before multiplying (by dividing 6 and 9 by 3):

$$\frac{6}{7} \times \frac{4}{9} = \frac{24}{63} = \frac{8}{21}$$

$$\frac{6}{7} \times \frac{4}{9} = \frac{2}{7} \times \frac{4}{3} = \frac{8}{21}$$

When you multiply a number by a proper fraction (a fraction less than 1), the result is smaller than the number.

Dividing proper fractions

Invert the second fraction to make its **reciprocal**. Then multiply.

This gives the same answer as the original division.

For example:

$$\frac{5}{8} \div \frac{3}{5} = \frac{5}{8} \times \frac{5}{3} = \frac{25}{24}$$

When you divide a number by a proper fraction (a fraction less than 1), the result is bigger than the number.

Exercise 2

1 **Calculate the answers to these multiplications. Give each answer as a proper or improper fraction in its simplest form.**

a) $\frac{4}{5} \times \frac{3}{8}$ b) $\frac{1}{9} \times \frac{2}{3}$ c) $\frac{5}{6} \times \frac{3}{10}$

....

d) $\frac{8}{11} \times \frac{1}{4}$ e) $\frac{5}{12} \times \frac{4}{15}$ f) $\frac{4}{5} \times \frac{4}{5}$

....

g) Explain why each answer is smaller than either of the numbers being multiplied.

....

....

2 Calculate the answers to these divisions. Give each answer as a proper or improper fraction in its simplest form.

a) $\frac{3}{4} \div \frac{1}{4}$ b) $\frac{1}{2} \div \frac{1}{8}$ c) $\frac{5}{9} \div \frac{3}{4}$

....

d) $\frac{2}{3} \div \frac{5}{6}$ e) $\frac{7}{12} \div \frac{2}{3}$ f) $\frac{3}{8} \div \frac{2}{7}$

....

g) Explain why each answer is bigger than the first number in the calculation.

3 Jarvis finds $\frac{3}{4} \times \frac{1}{6}$. His working is shown below.

$$\frac{3}{4} \times \frac{1}{6} = \frac{9}{12} \times \frac{2}{12} = \frac{18}{144} = \frac{1}{8}$$

a) Describe the unnecessary step that Jarvis has taken.

....

b) Show a more efficient way to do this calculation.

4 Saskia has $\frac{2}{3}$ of a packet of rice. She uses $\frac{3}{4}$ of the rice.

Calculate what fraction of a packet of rice she uses.

....

....

10 Manipulating expressions

You will practice how to:

- Understand how to manipulate algebraic expressions, including:
  - collecting like terms
  - applying the distributive law with a constant.

10.1 Simplifying expressions

Summary of key points

Expressions can be simplified by collecting like terms.

$$4r + 10 + 2r - 3 = 6r + 7$$

$$6a + 9b - a + 4b = 5a + 13b$$

Remember that $4r$ means $4 \times r$, and that the number is always written before the letter, so $b \times 9$ would be written as $9b$.

Exercise 1

1 Simplify:

a) $n + n + n = $ ....

b) $y + y + y + y + y = $ ....

c) $k + k + 2 = $ ....

d) $4m + 2m = $ ....

e) $5d + 3 + 2d + 1 = $ ....

f) $4f + 6 + 4f + 2 = $ ....

g) $2a + 3b + a + 4b = $ ....

h) $6t + u + 3t + 2 + u = $ ....

i) $4f + 6g + 3f + 1 + 3g + 2 = $ ....

j) $3m + 2n + 5m + 6 + n + 2m + 3 = $

....

2 Simplify:

a) $d + d - d = $ ....

b) $m + m + m + m - m = $ ....

c) $n - n + n - n = $ ....

d) $7t - 3t = $ ....

e) $8u - 2u = $ ....

f) $4g + 1 - 2g = $ ....

g) $5n - 2 + n = $ ....

h) $6h + 3 - 2h - 1 = $ ....

i) $6 + 6x + 8y - 2x + 3y = $ ....

j) $10a + 2b + 19 - a - b + 3 - b = $

....

3 Simplify:

a) $7 - 3n + 3 + 7n = $ ....

b) $4g + 11 - 9g - 2 = $ ....

c) $9 - 3m + 1 - 2m - 3 = $ ....

d) $3p + q + 1 - p - 3q + 4 = $ ....

e) $6 - 3d + e + 1 - 2d - 6e = $ ....

f) $7m^2 - 11 + m - 3m^2 + m - 2 = $ ....

4 Draw lines to match equivalent expressions.

$4p - 3q + 2p + q$		$2p - 2q$

$q - 3p - 3q + 5p$		$6p - 2q$

$11p - 4q - 5p + 8q$		$2p + 4q$

$5p - 3q - 3p + 7q$		$6p + 4q$

5 Simplify:

a) $5 \times a = $ ....

b) $4 \times y = $ ....

c) $t \times 7 = $ ....

d) $2m \times 4 = $ ....

e) $7 \times 2w = $ ....

f) $4p \times 8 = $ ....

g) $3 \times y \times 2 = $ ....

h) $4g \times 2 \times 5 = $ ....

6 Form and simplify an expression, in cm, for the perimeter of each shape.

a)

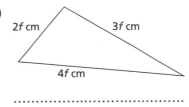

$2f$ cm $3f$ cm

$4f$ cm

....

.... cm

b)

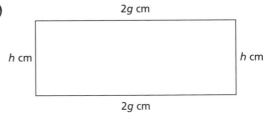

$2g$ cm

h cm h cm

$2g$ cm

....

.... cm

c)

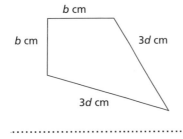

b cm

b cm $3d$ cm

$3d$ cm

....

.... cm

d)

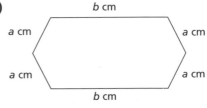

b cm

a cm a cm

a cm a cm

b cm

....

.... cm

7 The perimeter of a rectangle is $2c + 4d$ cm. The length of one side is c cm. What is the length of the other side?

....cm

8 Match an expression from the first column with one from the second column.

$\frac{x}{3} + \frac{x}{3}$
$\frac{x}{5} + \frac{3x}{5}$
$\frac{x}{3} + \frac{x}{6}$
$\frac{4x}{5} - \frac{x}{5}$
$\frac{5x}{6} - \frac{x}{2}$

$\frac{x}{2}$
$\frac{3x}{5}$
$\frac{2x}{3}$
$\frac{x}{3}$
$\frac{4x}{5}$

Think about

9 Find 10 expressions that simplify to $12w + 8$.

10.2 Multiplying a constant over a bracket

Summary of key points

To **expand** or **multiply out** a bracket, you multiply all the terms inside the bracket by the number (constant) on the outside.

$$4(n + 3) = 4n + 12$$

$$5(2x + 5y) = 10x + 25y$$

Exercise 2

1 Expand the brackets.

a) $4(n + 3) =$ ....

b) $7(x - 1) =$ ....

c) $3(y - 4) =$ ....

d) $2(f + g + 5) =$ ....

e) $4(2d + 7) =$ ....

f) $9(4t - 5) =$ ....

g) $4(2u + 3v - 2) =$ ....

h) $5(6w - 7x) =$ ....

i) $7(1 + 10m) =$ ....

j) $3(2y + 5z - 3x) =$ ....

2 Complete these statements.

a) $6(t + \text{........}) = 6t + 24$

b) $2(\text{........} - 8) = 6y - 16$

c) $5(\text{........} + 4) = 25g + \text{........}$

d) $\text{........}(3e + 4f) = 18e + \text{........}$

3 Expand and simplify:

a) $2(m + 3) + 3(m + 4) =$

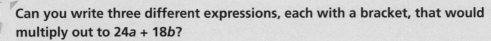

b) Show that

$4(2x + 1) + 5(x + 2) = 13x + 14$

Think about

4 **Can you write three different expressions, each with a bracket, that would multiply out to 24a + 18b?**

What do you notice about the numbers outside your brackets?

11 Angles

You will practice how to:

- Know that the sum of the angles around a point is 360°, and use this to calculate missing angles.
- Derive the property that the sum of the angles of a quadrilateral is 360°, and use this to calculate missing angles.
- Recognise the properties of angles on: parallel lines and transversals, perpendicular lines and intersecting lines.

11.1 Angles around a point

Summary of key points

Angles on a straight line add up to 180°.

$a + b = 180°$

Angles around a point add up to 360°.

$a + b + c = 360°$

Vertically opposite angles are equal.

$a = b$

Exercise 1

1 Work out the size of each lettered angle.

a)

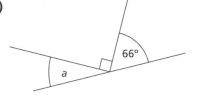

b)

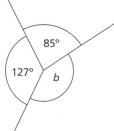

66°

85°   127°   b

a = ...°   b = ...°

c)

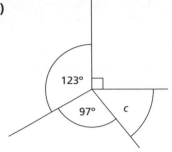

d)

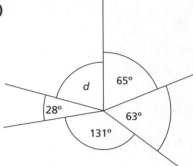

$c =$ ....° $d =$ ....°

2 Work out the value of x in each diagram.

a)

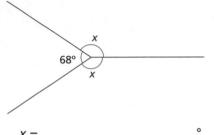

$x =$ ....°

b)

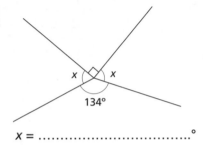

$x =$ ....°

c)

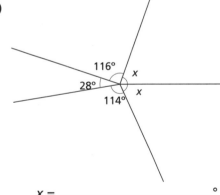

$x =$ ....°

3 Find the size of the marked angles on each diagram.

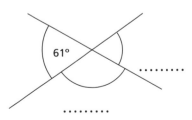

4 **a)** Angelique has three angles that fit around a point.

Write down a possible set of values for Angelique's three angles.

........................° , ...° and ....°

b) Henri draws this diagram showing three angles making a straight line.

68° 77° 45°

Give a reason why Henri's diagram must contain an error.

....

....

5 Ben faces the school gate.

He turns clockwise through 100° to face a shop.

He then turns a further 60° to face a tree.

He then turns a further 120° to face a lorry.

a) Complete the sketch showing possible positions of the shop, the tree and the lorry

School gate
×

×
Ben

b) By how much does Ben now need to turn clockwise so that he faces the school gate again?

........................°

6 Kadeem draws a diagram with three angles around a point. Two of the angles are twice the size of the other one.

a) Draw a sketch of the diagram. **b)** Calculate the size of all three angles.

....

Summary of key points

Angles in a triangle add up to 180°.

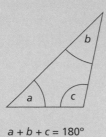

$a + b + c = 180°$

Angles in a quadrilateral add up to 360°.

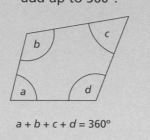

$a + b + c + d = 360°$

Exercise 2

1 **Here are six quadrilaterals.**

Quadrilateral *A*

Quadrilateral *B*

Quadrilateral *C*

120°

x

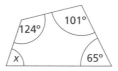

124° 101°

x 65°

101°
109°

x

Quadrilateral *D*

Quadrilateral *E*

Quadrilateral *F*

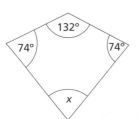

132°

74° 74°

x

76°
127°

87°

x

131°
49°
32°

x

Put each quadrilateral in the correct column of the table by finding each value of *x*.

The first one has been done for you.

x = 50°	*x* = 60°	*x* = 70°	*x* = 80°
	A		

2 Find the size of each lettered angle.

a)

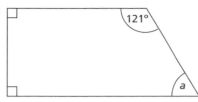

a = ...°

b)

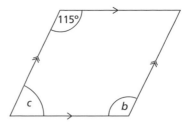

b = ...°   c = ...°

c)

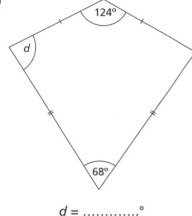

d = ...°

3 ABCD is a quadrilateral. E is a point on AD.

Complete the following statements.

a) Angle DEC = ...° because
angles on a line add up to ....°.

b) Angle ECD = ...° because angles
in a triangle add up to ....°.

c) Angle BCE = ...° because angles
in a quadrilateral add up to ....°.

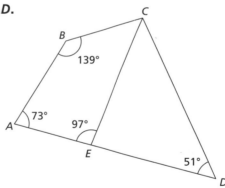

4 Complete this proof about the sum of the angles in a quadrilateral.

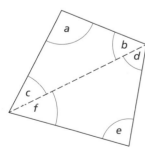

The angles in a triangle add up to ....°.

So   a + b + c = ...°

and   d + e + ... = ...°

So   a + b + c + d + e + f = ...°

a + (b + d) + e + (............ + ...) =
............°

So the four angles in the quadrilateral add up

to ....°.

5 Show that $x = 45°$

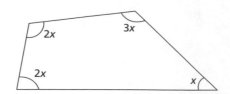

Think about

6 Make up your own angle problem involving quadrilaterals. The solution to your problem should be **142°**.

11.3 Angles and parallel lines

Summary of key points

A **transversal** is a line that crosses two or more lines.

Equal angles are formed when a transversal crosses parallel lines.

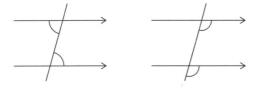

These angles are equal   These angles are equal

Exercise 3

1 Colour in red all the angles that are equal to the marked angle. Use tracing paper or a protractor to help you.

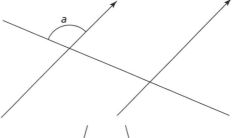

2 a) Colour in blue all the angles that are equal to the angle marked *b*.

   b) Colour in green all the angles equal to the angle marked *c*.

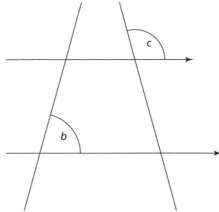

---

3 Write whether the marked angles are equal or not equal.

a)
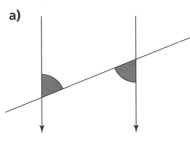

b)
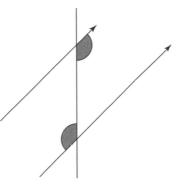

c)
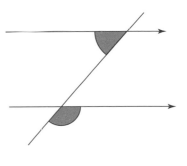

....

d)
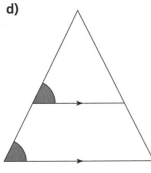

e)
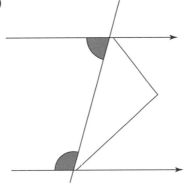

f)
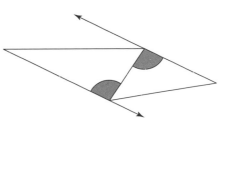

....

4 Write the size of the angle marked by the letter. The diagrams are not to scale.

a)
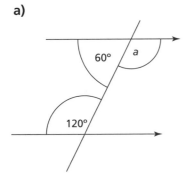

b)
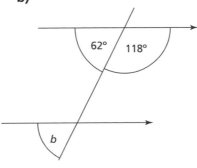

c)
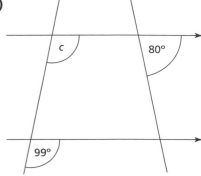

a = ...°

b = ...°

c = ...°

d)

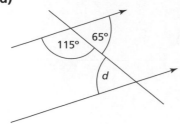

e)

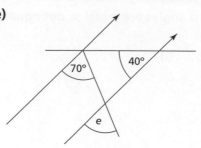

f)

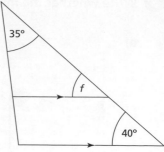

d = ...°

e = ...°

f = ...°

5 Label the diagram with all the angles you can work out.

The diagram is not to scale.

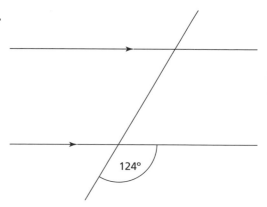

6 Omar says that lines *AB* and *CD* are parallel.

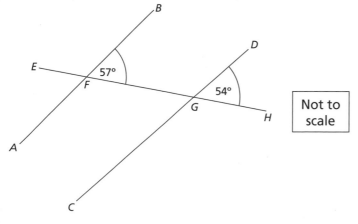

Not to scale

Explain why Omar is wrong.

....

....

Measures of average and spread

You will practice how to:

- Use knowledge of mode, median, mean and range to describe and summarise large data sets. Choose and explain which one is the most appropriate for the context.
- Interpret data, identifying patterns, within and between data sets, to answer statistical questions. Discuss conclusions, considering the sources of variation, including sampling, and check predictions.

12.1 Mean, median, mode and range

Summary of key points

The **mean**, **median** and **mode** are all types of **average**.

The **mean** is the total value of all the data divided by the number of pieces of data.

The **mode** (or the **modal value**) is the value in a set of data that occurs most frequently.

A set of data can have more than one mode (or no mode at all).

The **median** is the middle value of data, once the data is arranged in order of size.

When there are n values in a data set, we use the expression $\frac{n+1}{2}$ to find the position of the median.

If $\frac{n+1}{2}$ is not a whole number, the median will be halfway between the middle pair of values.

The **range** is a measure of spread.

The range is the difference between the highest data value and the lowest data value.

It shows how consistent the data is.

Exercise 1

 1 **Find the modal number of pets.**

Number of pets

2	1	0	1	3	5	2
0	0	4	9	6	1	4

Mode ....

2 **Find the median and range of each set of data.**

Put the data in order first.

a) Temperature (in °C) of 13 cities:

21   23   22   19   29   25   23   22   18   17   18   20   22

Median = ...°C   Range = ...°C

b) Speed (in km/h) of 12 cars:

42   38   51   40   55   36   43   50   42   58   47   54

Median = ... km/h Range = ... km/h

3 **The lengths, to the nearest tenth of a millimetre, of 15 insects are:**

14.6   13.2   12.8   15.3   19.1

17.5   13.5   14.8   18.4   20.0

17.3   17.6   18.1   16.7   14.4

a) Find the median length of the insects.

.... mm

b) Calculate the mean length of the insects.

.... mm

c) Explain why it is not possible to find the mode.

....

....

4 **The temperatures, in °C, of 16 cities are:**

7.3   4.2   0.9   10.4   3.5   2.8   9.4   7.9

5.2   6.7   9.6   13.5   8.2   4.3   11.2   3.7

a) Find the median temperature.

.... °C

b) Calculate the mean temperature.

.... °C

5 Write six numbers which have:

a) a mode of 4

....

b) a mode of 7 and a range of 3

....

6 Noor records the heights (in cm) of some plants.

25 19 20 22 17 23 20 21 19 28 24 20 21 26 24 23

Tick to show if each statement is true or false.

	True	False
a) The modal height is 20 cm.	☐	☐
b) The median height is 20 cm.	☐	☐
c) The range is 9 cm.	☐	☐

Match each set of numbers to its mean.

11 14 17 12 11		Mean = 10
10 12 11 16 11 12		Mean = 11
15 10 4 12 13 11 12		Mean = 12
13 3 16 4 12 10 11 11		Mean = 13
12 9 14 16 19 11 17 15 13		Mean = 14

A family consists of Jaleel, Fatima and their three children.

The total of the ages of the children is 24 years.
Jaleel is 34 years.
Fatima is 32 years.
Find the mean age of the five people in the family.

....

.... years

9 Maria records the heights of 12 young children. The mean height is 76 cm. Find the sum of the heights of the children.

....

.... cm

10 The mean mass of six biscuits is 23 g. The masses of five of the biscuits are:

20 g   19 g   25 g   23 g   24 g

Work out the mass of the sixth biscuit.

.... g

11 Graham and India each think of a set of four whole numbers.

3   6   7   12
Graham's numbers

6   13   ?   ?
India's numbers

The mean of India's numbers is 2 more than the mean of Graham's numbers.
The range of India's numbers is equal to the range of Graham's numbers.
Find the two missing numbers.

....

....

.... and ....

Think about

12 In what situations is it sensible to use the mode as an average?

Summary of key points

Median and mode from a simple table

The table shows the ages of some children.

Age (years)	Frequency
7	2
8	4
9	3
10	1

The modal age is 8 years (as this has the highest frequency).

There are 10 children, so the median is halfway between the ages of the 5th and 6th children. These are both 8 years, so the median is also 8 years.

Mean from a simple table

The mean can be found by adding an extra column to the table.

Age (years)	Frequency	Age × frequency
7	2	14
8	4	32
9	3	27
10	1	10
TOTALS	**10**	**83**

The mean can be found by dividing the column totals.

Mean = $\frac{83}{10}$ = 8.3 years

Exercise 2

1. Here are three frequency tables. Find the median, mode and range for the data in each table.

Table A

Value	Frequency
10	1
11	0
12	5
13	3
14	1

Table B

Value	Frequency
10	1
11	7
12	1
13	0
14	1

Table C

Value	Frequency
10	0
11	1
12	2
13	3
14	4

	Median	Mode	Range
Table A			
Table B			
Table C			

2 The number of students in 15 classes in a school is shown in the table.

Number of students in a class	Frequency
28	5
29	4
30	3
31	1
32	2

Find the median, mode and range for the number of students in the class.

Median......... Mode......... Range.........

3 The frequency table shows the number of people living in 14 houses.

Number of people living in house	Number of houses	Number of people × number of houses
1	2	2
2	4	
3	2	
4	4	
5	2	
TOTAL	14	

a) Complete the table.

b) Find the mean number of people in a house.

....

....

....

4 The frequency table shows the number of instruments the children in a class play.

Number of instruments	Number of children	Number of instruments × number of children
0	4	
1	8	
2	6	
3	2	
TOTAL		

a) Complete the table.

b) How many children are there in the class? ....

c) Find the mean number of instruments played. ....

....

5 The frequency table shows the time (to the nearest second) that a group of children took to run a race.

Time (seconds)	Frequency
23	4
24	8
25	10
26	11
27	6
28	1

Find the mean time for the race. .... seconds

....
....

6 The table shows the number of trees growing in each of 60 gardens.

Number of trees, n	Frequency, f
0	18
1	24
2	10
3	2
4	6
TOTAL	

a) Write down the modal number of trees. ....

b) Calculate the mean number of trees per garden.
Give your answer correct to 2 decimal places. ....

c) Pat says that the range is 24 − 2 = 22. Explain why he is wrong.

....

....

7 The table shows the number of staff working in an office on 40 different days.

Number of staff	Frequency
11	3
12	7
13	11
14	9
15	8
16	2

Seb says that the mean number of staff is greater than the mode. Is he correct?
Show your working.

....

....

....

....

8 A fruit wholesaler shop claims that the mean number of oranges in a box is 53.

Mani buys 25 boxes to sell the oranges in his shop. He counts the number oranges in each box and produces the table below.

Number of oranges	Frequency
50	1
51	3
52	5
53	8
54	4
55	3
56	1

a) Is the wholesaler's claim about the mean number of oranges in each box accurate? Give reasons for your answer.

....

....

....

b) Another box is added to the table. It contains 57 oranges. Explain what effect this will have on the mean, median, mode and range.

Mean

....

....

Median

....

....

Mode

....

....

Range

....

....

13 Calculations

You will practice how to:

- Understand that brackets, positive indices and operations follow a particular order.
- Use knowledge of common factors, laws of arithmetic and order of operations to simplify calculations containing decimals or fractions.

13.1 Order of operations

Summary of key points

In calculations involving several operations, the order to work them out is:

1. Brackets

2. Indices (powers and roots)

3. Multiplication and division (from left to right)

4. Addition and subtraction (from left to right)

When working out calculations inside a bracket, follow the usual order of operations.

If there are brackets inside brackets, find the calculation in the innermost brackets first, then work outwards.

Exercise 1

1. Calculate:

   a) $14 - 4 \times 3$ .... **b)** $20 + 2 - 4$ ....

   c) $5 + 15 \div 5$ .... **d)** $16 \div 4 \times 2$ ....

2. Write the missing number in each calculation.

   .... $\times (14 - 5) = 27$

   $25 - (10 + ...) = 7$

   $8 \times (20 - ....) = 88$

   $3 + 9 \times .... = 48$

   $40 \div (........ \div 2) = 5$

   $3 \times 7 + 6 \times .... = 57$

3. Calculate:

   a) $120 \div 4 + 2 \times 3$ .... **b)** $120 \div 4 + (2 \times 3)$ ....

   c) $120 \div (4 + 2 \times 3)$ .... **d)** $120 \div (4 + 2) \times 3$ ....

   e) $(120 \div 4 + 2) \times 3$ .... **f)** $(120 \div 4) + 2 \times 3$ ....

4 Here is Rebecca's homework.

a.	$3 \times (8-3) = 24 - 3 = 21$	b.	$(5-2)^2 = 25 - 4 = 21$
c.	$21 - 3 \times 2 + 1 = 21 - 3 \times 3 = 12$	d.	$(29+6) \div (11-4) = 35 \div 7 = 5$
e.	$4 + 20 \div 2^2 = 4 + 20 \div 4 = 4 + 5 = 9$	f.	$((5-2) \times 3)^2 = (5-2) \times 9 = 27$

Mark each of her answers as correct or incorrect.

For each incorrect answer, write the correct answer underneath.

5 Add one pair of brackets to make each calculation correct.

a) $20 + 6 + 4 \div 2 = 25$

b) $6 \times 3 + 2 \times 2 = 40$

c) $6 \div 2 \times 3 + 2^2 = 5$

d) $5 - 2^2 + 10 \div 5 + 5 = 16$

Think about

6 Make up 5 calculations of your own with answer 21. Each calculation should involve a power and at least two operations (+, –, × or ÷).

13.2 Simplifying calculations with decimals and fractions

Summary of key points

You can rewrite some calculations to make them easier to do, without changing the results. The table shows some methods you can use.

Method	Example
changing the order of numbers in addition (addition is **commutative**)	$0.8 + 0.36 + 0.2 = 0.8 + 0.2 + 0.36 = 1 + 0.36$
changing the order of numbers in multiplication (multiplication is commutative)	$17 \times 20 \times \frac{1}{2} = \frac{1}{2} \times 20 \times 17 = 10 \times 17$
changing the order of additions and subtractions (if there are no other types of operation in the calculation)	$1.45 + 3.87 - 0.45 = 1.45 - 0.45 + 3.87 = 1 + 3.87$
changing the order of multiplications and divisions (if there are no other types of operation in the calculation)	$28 \times 9 \div 7 = 28 \div 7 \times 9 = 4 \times 9$
using the **distributive law** to rewrite a multiplication by **partitioning** a number	$\mathbf{11} \times 27 = \mathbf{10} \times 27 + \mathbf{1} \times 27 = 270 + 27$
using the distributive law in reverse to rewrite a multiplication	$1.7 \times 8 + 1.7 \times 2 = 1.7 \times 10$

1 a) Complete the calculation ...... + 8.3 − ...... = 8.3 in two different ways.

....

b) Complete the calculation ...... × 5.2 − ...... × 5.2 = 52 in two different ways.

....

2 Find:

a) 4 × 11 × 25 ....   b) 4 × 10.7 × 2.5 ....

c) 6.5 + 13.4 − 3.5 ....   d) 0.72 + 1.95 + 0.05 ....

e) $\frac{8}{9} \times 81$ ....   f) $\frac{5}{6} + 6\frac{7}{8} + \frac{1}{6}$ ....

g) 300 × 19.5 ÷ 3 ....   h) 72 × 0.6 ÷ 9 ....

3 Tick (✓) the calculations that give the same answer as 91 × 3.7 + 9 × 3.7

A 3.7 × 81 + 3.7 × 19 ☐   B 65 × 3.7 + 25 × 3.7 ☐

C 100 × 3.7 ☐   D 3.7 × 9 + 81 × 3.7 ☐

4 Find:

a) 97 × 4.3 + 3 × 4.3 ....   b) 21.6 × 6 + 21.6 × 4 ....

c) 27 × 1.1 ....   d) $\frac{8}{11} \times 22$ ....

5 Tick (✓) the calculations that are equivalent to $\frac{15}{8} \times \frac{24}{5}$.

A 15 × 24 ÷ 8 × 5 ☐   B 15 ÷ 8 × 24 ÷ 5 ☐

C 15 ÷ 5 × 24 ÷ 8 ☐   D 24 ÷ 8 × 15 ÷ 5 ☐

6 Jared does three calculations. His working is shown below.

a) 700 × 4.6 ÷ 7 = 3220 ÷ 7 = 460

b) $\frac{4}{11} + \frac{3}{7} + \frac{7}{11} = \frac{28}{77} + \frac{33}{77} + \frac{49}{77} = \frac{110}{77} = \frac{10}{7} = 1\frac{3}{7}$

c) 7 × 3.6 + 3.6 × 3 = 25.2 + 10.8 = 36

Show an easier method for each calculation.

a) ....   b) ....   c) ....

....

....

14 Functions and formulae

You will practice how to:

- Understand that a function is a relationship where each input has a single output. Generate outputs from a given function and identify inputs from a given output by considering inverse operations (linear and integers).
- Understand that a situation can be represented either in words or as a formula (single operation), and move between the two representations.

14.1 Functions and mappings

Summary of key points

A **function** is a mathematical process that gives an output value for a given input value.

You can show a function using a **function machine**, a **mapping** or a **formula**.

An **inverse function** undoes what the original function did.

Exercise 1

1 Here is a function machine.

Complete this table.

Input	Output
2	10
4	
6	
20	

2 Here is a function machine.

INPUT ⟶ ☐ + 2 ⟶ OUTPUT

a) Find the missing numbers.

0 ↦ ....

3 ↦ ....

4 ↦ ....

b) Complete the mapping diagram to show the answers to part a).

0 1 2 3 4 5 6

0 1 2 3 4 5 6

3 a) Complete the table for the function $x \mapsto x + 4$

Input	Output
1	5
3	
	10
10	
20	

b) If the output for this function was 100, what was your input number?

....

4 Match the function machines to their formulae.

input → ☐ × 6 → output $y = x - 6$

input → ☐ + 8 → output $y = 8x$

input → ☐ + 6 → output $y = x + 8$

input → ☐ × 8 → output $y = \dfrac{x}{6}$

input → ☐ − 6 → output $y = 6x$

input → ☐ ÷ 6 → output $y = x - 8$

input → ☐ − 8 → output $y = x + 6$

input → ☐ ÷ 8 → output $y = \dfrac{x}{8}$

5 Two of these functions have the same output value when the input is 3.
Which one is the odd one out?

Function machine A $3 \rightarrow$ $\boxed{\times 5}$ $\rightarrow output$

Function machine B $3 \rightarrow$ $\boxed{squared}$ $\rightarrow output$

Function machine C $3 \rightarrow$ $\boxed{+12}$ $\rightarrow output$

Function machine ....

6 For each of these mappings, draw the function machine and the inverse
function machine.

Mapping	$x \mapsto x + 7$	$x \mapsto 4x$
Function machine		
Inverse function machine		

7 For each of these functions, write the inverse function.

a) $x \mapsto x + 5$ ....

b) $x \mapsto 6x$ ....

c) $x \mapsto x - 8$ ....

d) $x \mapsto \dfrac{x}{5}$ ....

8 The function $x \mapsto 2x$ will always give an even answer output when a positive
whole number is input.

a) Find a function that will always give an odd number output when a positive
whole number is input.

....

b) Use your function to complete the table of values.

Input	Output
1	
10	
21	
38	
50	

9 How many different function machines can you find that map 3 to 12?

14.2 Constructing and using formulae

Summary of key points

There are seven days in one week.

A word **formula** to find the number of days in any number of weeks is:

number of days (d) = 7 × number of weeks (w)

This can be written as:

$d = 7w$

In 13 weeks, the number of days is:

$d = 7 \times 13 = 91$

Remember that a formula is a mathematical relationship between two or more variables expressed **algebraically.**

A **variable** is a letter that represents a number that can take different values.

Exercise 2

1 Nadja is a decorator. She uses this formula to work out how much to charge a customer.

charge (in dollars) = number of hours worked × 15

Use Nadja's formula to find how much she charges a customer if she works for:

a) 3 hours $ ....   **b)** 11 hours $ ....

2 Oscar makes cakes. He uses this formula to find how many eggs he needs.

number of eggs needed = number of cakes made × 4

Use Oscar's formula to find how many eggs he needs if he makes seven cakes. ....

3 Circle the correct formula for the number of metres, y, in x kilometres.

$y = 1000 + x$ $y = 1000x$ $y = 1000 - x$ $y = \dfrac{1000}{x}$

4 a) Write a formula for the number of minutes, m, in h hours. ....

b) Use your formula to find the number of seconds in 4 hours.

.... minutes

5 a) Write a formula for the number of millilitres, m, in c centimetres. ....

b) Use your formula to find the number of millilitres in 7.8 centimetres.

.... millilitres

6 A packet contains six cakes.

a) Write a formula for the number of cakes, c, in p packets. ....

b) Use your formula to find the number of cakes in 9 packets. ....

7 A shop sells books on the internet. The total cost it charges to a customer is found by adding a delivery charge of $3 to the cost of the books ordered.

a) Write a formula to find the total cost, T, for

a customer who orders books that cost $b. ....

b) Use your formula to find the total cost for

a customer who orders books costing $11. $ ....

8 A cinema works out the cost, c, of a child ticket by subtracting $2 from the cost, a, of an adult ticket.

Write a formula for c in terms of a. ....

9 a) Fiona earns $12 an hour. Write a formula for the total

amount, F, that Fiona earns if she works for n hours. ....

b) Marco earns $h an hour. Write a formula for the total

amount, M, that Marco earns if he works for n hours. ....

10 The total cost, C, of buying n chairs in a furniture shop is given by the formula $C = 60n$.

A table costs $140.

Celia has $400.

Can she afford to buy a table and six chairs? Explain your answer.

....

....

....

15 Area and volume

You will practice how to:

- Understand the relationships and convert between metric units of area, including hectares (ha), square metres (m²), square centimetres (cm²) and square millimetres (mm²).
- Derive and know the formula for the area of a triangle. Use the formula to calculate the area of triangles and compound shapes made from rectangles and triangles.
- Derive and use a formula for the volume of a cube or cuboid. Use the formula to calculate the volume of compound shapes made from cuboids, in cubic metres (m³), cubic centimetres (cm³) and cubic millimetres (mm³).
- Use knowledge of area, and properties of cubes and cuboids, to calculate their surface area.

15.1 Converting between units of area

Summary of key points

$1\ cm^2 = 100\ mm^2$ $\qquad$ $1\ m^2 = 10\ 000\ cm^2$ $\qquad$ $1\ km^2 = 1\ 000\ 000\ m^2$

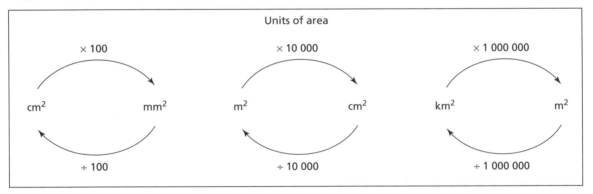

Units of area

Large areas (such as the area of a field or a park) can be measured in **hectares**.

1 hectare (ha) $= 10\ 000\ m^2$

Exercise 1

1 Complete the tables to show the correct conversions between units.

Area in cm²	Area in mm²
8 cm²	800 mm²
45 cm²	.... mm²
.... cm²	150 mm²
.... cm²	67 mm²

Area in m²	Area in cm²
3 m²	30 000 cm²
7.15 m²	.... cm²
.... m²	87 600 cm²
.... m²	6840 cm²

2 Complete these conversions.

a) 0.5 ha = ... m²

b) 3 ha = ... m²

c) 600 m² = ... ha

d) 24 000 m² = ... ha

3 Write this set of areas in order of size, starting with the smallest.

0.08 m²	42 000 cm²	650 000 m²	95 000 mm²	0.5 km²
....	....	....	....	....

smallest   largest

4 Put a ring around the larger area in each pair.

a) 6.5 hectares   or   6800 m²

b) 0.34 hectares   or   3420 m²

c) 77 000 m²   or   7.75 hectares

d) 11 100 m²   or   1.07 hectares

5 A piece of card has an area of 600 cm². An area of size 450 mm² is removed from the card.
Calculate the area of card that remains.

.... cm²

6 Convert 0.0084 m² to square millimetres.

.... mm²

7 Jan has a farm. Every 1000 m² of field yields 25 bales of hay.
Find the yield of the 40 hectares of field on Jan's farm.

....

8 A tennis court is 24 m by 11 m.
Alex said 'You can fit 38 tennis courts in a hectare.'
Is he correct? Explain your answer.

....

....

Think about

9 Find the approximate area in hectares of a football pitch.

Summary of key points

The area of a triangle is found by:

Area = $\frac{1}{2}$ × base × height   or $A = \frac{1}{2}bh$

Example:

The area of this triangle is $\frac{1}{2}$ × 14 × 5 = 35 m²

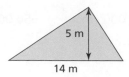

5 m

14 m

Exercise 2

1 **Calculate the areas of these compound shapes.**

The shapes in this exercise are not drawn to scale.

a)

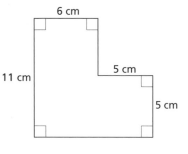

6 cm

11 cm

5 cm

5 cm

....

.... cm²

b)

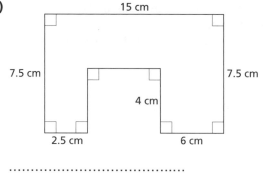

15 cm

7.5 cm

7.5 cm

4 cm

2.5 cm

6 cm

....

.... cm²

2 **Find the areas of triangles A–F. Then write each one in the correct column of the table.**
One has been done for you.

Triangle A

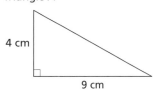

4 cm

9 cm

Triangle B

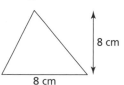

8 cm

8 cm

Triangle C

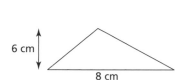

6 cm

8 cm

Triangle D

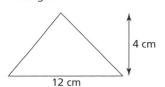

4 cm

12 cm

Triangle E

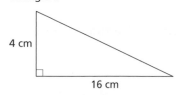

4 cm

16 cm

Triangle F

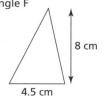

8 cm

4.5 cm

Area = 18 cm²	Area = 24 cm²	Area = 32 cm²
A		

3 Calculate the area of each shape. Then put a ring around the odd one out in each set.

a)

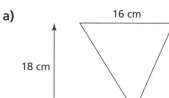 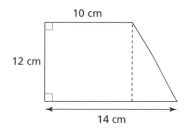 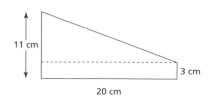

.... cm²   .... cm²   .... cm²

b)

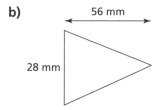 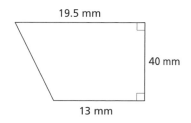 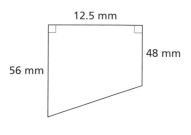

.... mm²   .... mm²   .... mm²

4 The area of the rectangle is the same as the area of the triangle. Calculate the value of *h*.

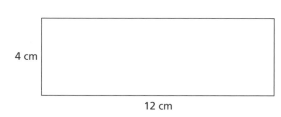 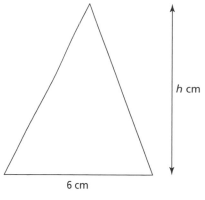

h = ...

5 Calculate the shaded area of this triangle.

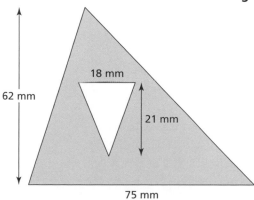

.... mm²

6 Calculate the area of this compound shape.

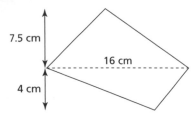

.... cm²

7 A triangle has an area of 0.45 cm². The base of the triangle is 9 mm. Calculate the height.

.... mm

15.3 Volume of cuboids

Summary of key points

Some volumes can be found by counting cubes.

This cuboid is made from centimetre cubes.

There are 3 layers with 12 cubes in each layer.

Its volume is 36 cm³.

There is a formula for finding the volume of a cuboid. This formula is usually more useful.

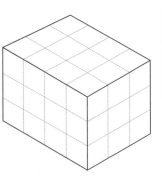

The volume of a shape is how much space it occupies. Volume is measured in cubed units, such as cm³ or mm³ or m³.

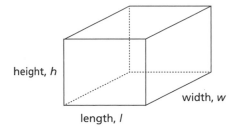

Volume of cuboid = length × width × height

1 These cuboids are made from centimetre cubes. Find the volume of each cuboid.

Diagrams are not drawn to scale.

a)

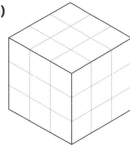

Volume = ... cm³

b)

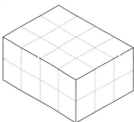

Volume = ... cm³

c)

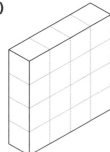

Volume = ... cm³

d)

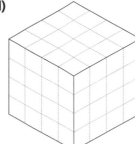

Volume = ... cm³

2 Sam makes four cuboids, each using exactly 60 centimetre cubes. The diagrams show the bottom layer of each of her cuboids. Find how many layers each cube must have.

Cuboid 1

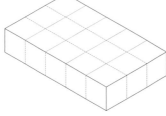

.... layers

Cuboid 2

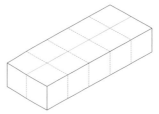

.... layers

Cuboid 3

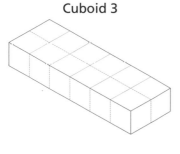

.... layers

Cuboid 4

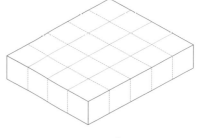

.... layers

3 Write these cuboids in order of volume, smallest first.

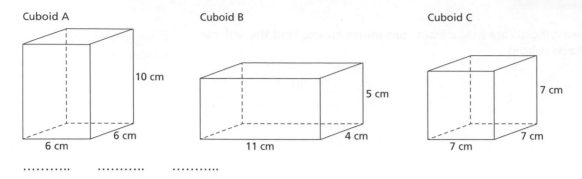

Cuboid A

10 cm

6 cm   6 cm

6 cm

Cuboid B

5 cm

11 cm   4 cm

Cuboid C

7 cm

7 cm   7 cm

7 cm

....

4 Circle the cuboids with a volume equal to 480 cm².

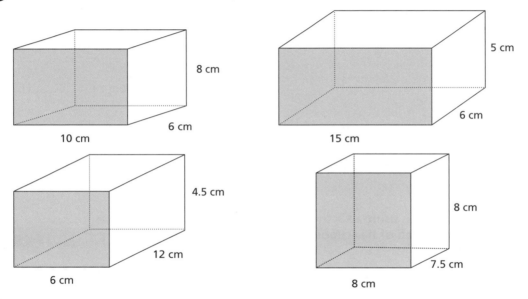

8 cm

6 cm

10 cm

5 cm

6 cm

15 cm

4.5 cm

12 cm

6 cm

8 cm

7.5 cm

8 cm

5 Find the difference in volume between each pair of cuboids.

a)

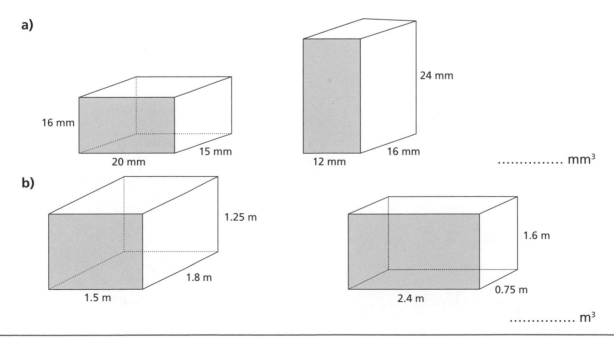

16 mm

20 mm   15 mm

24 mm

12 mm   16 mm

.... mm³

b)

1.25 m

1.5 m   1.8 m

1.6 m

2.4 m   0.75 m

.... m³

6 Tick the correct volume for each prism.

280 cm³   300 cm³

a)

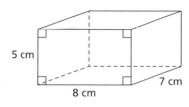

5 cm
8 cm
7 cm

☐   ☐

b)

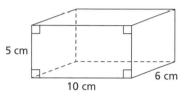

5 cm
10 cm
6 cm

☐   ☐

c)

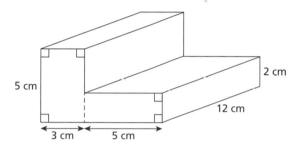

5 cm
2 cm
12 cm
3 cm   5 cm

☐   ☐

7 A solid shape is formed by joining two cuboids as shown. Calculate the total volume of the shape.

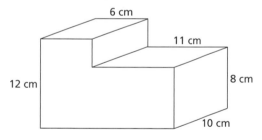

6 cm
11 cm
12 cm   8 cm
10 cm

Volume = ……….. cm³

8 The two cuboids have equal volumes. Find the value of *x*.

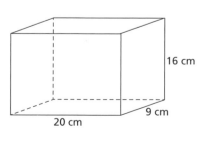

16 cm
20 cm   9 cm

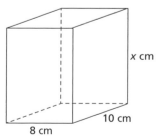

x cm
8 cm   10 cm

x = ………..

9 This prism is made from two cuboids. The volume of the prism is 360 cm³.

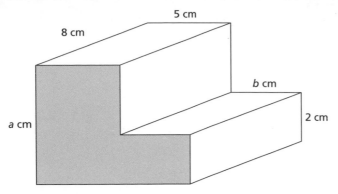

Find possible values for *a* and *b*.

$$a = \text{.....................} \quad b = \text{.....................}$$

Summary of key points

A cuboid has six rectangular faces.

A cube has six square faces.

The surface area of a cuboid is the total area of all of its faces.

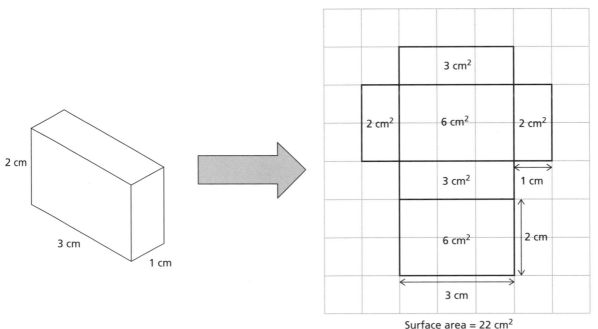

Surface area = 22 cm²

1 **Complete the net of each cuboid by drawing in the missing face.**

a)

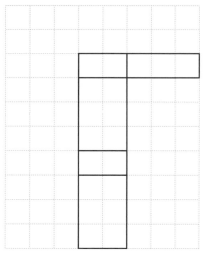

b)

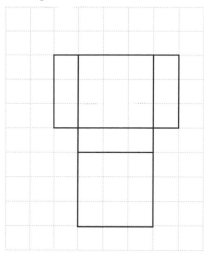

2 **The nets of two cuboids are shown. For each cuboid:**

- mark on the missing measurements
- find the surface area.

a)

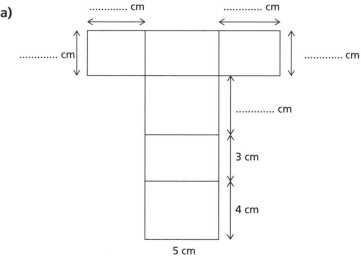

.... cm

.... cm

.... cm

.... cm

3 cm

4 cm

5 cm

.... cm²

b)

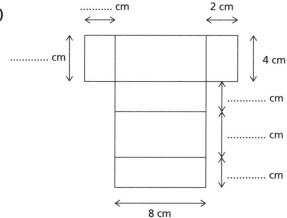

.... cm

2 cm

.... cm

4 cm

.... cm

.... cm

.... cm

8 cm

.... cm²

3 The diagram shows the net of a cuboid.

a) Find the missing measurements.

b) Find the surface area.

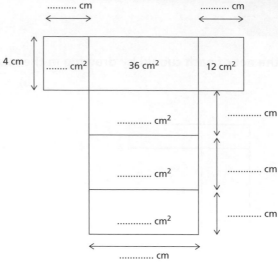

.... cm   .... cm

4 cm

.... cm²   36 cm²   12 cm²

.... cm²

.... cm

.... cm²

.... cm

.... cm²

.... cm

.... cm

.... cm²

4 By sketching a net, find the surface area of each cuboid.

a)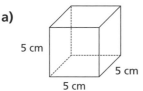

5 cm

5 cm

5 cm

.... cm²

b)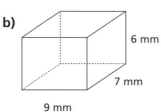

6 mm

7 mm

9 mm

.... mm²

5 Circle the two cuboids that have the same surface area.

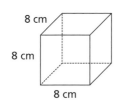

8 cm

8 cm

8 cm

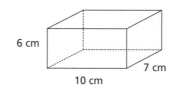

6 cm

7 cm

10 cm

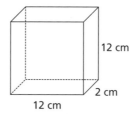

12 cm

2 cm

12 cm

6 A cuboid has four rectangular faces and two square faces. Each rectangular face measures 8 cm by 6 cm. Find the two possible values for the surface area of the cuboid.

.... cm²   or .... cm²

Think about

7 Minal has 400 cubes. First she uses the cubes to form a cuboid measuring 8 cubes by 5 cubes by 10 cubes.

Find a way to arrange all the cubes into a cuboid with a larger surface area.

16 Fractions, decimals and percentages

You will practice how to:

- Recognise that fractions, terminating decimals and percentages have equivalent values.
- Understand the relative size of quantities to compare and order decimals and fractions, using the symbols =, ≠, > and <.

16.1 Equivalent fractions, decimals and percentages 1

Summary of key points

To convert a **terminating decimal** to the **equivalent** percentage, multiply by 100:

0.08 = 8%   0.7 = 70%   0.36 = 36%

To convert a percentage to the equivalent decimal, divide by 100:

99% = 0.99   60% = 0.6 (or 0.60)

To convert a terminating decimal to the equivalent fraction, use the smallest place value in the decimal as the denominator. To find the numerator, multiply the decimal by the denominator value. Then simplify the fraction if possible.

The smallest place value in 0.9 is tenths, so $0.9 = \dfrac{9}{10}$

The smallest place value in 0.28 is hundredths, so $0.28 = \dfrac{28}{100} \overset{\text{divide by 4}}{\underset{\text{divide by 4}}{\rightarrow}} \dfrac{7}{25}$

Some fractions can be written with a denominator of 10 or 100. You can convert these to terminating decimals or percentages:

$\dfrac{3}{5} = \dfrac{6}{10} = 0.6 = 60\%$ $\dfrac{7}{100} = 0.07 = 7\%$ $\dfrac{9}{25} = \dfrac{36}{100} = 0.36 = 36\%$

Exercise 1

1 Choose from the box the decimal that is equivalent to each of these percentages.

a) 24% ....   b) 47% ....

c) 15% ....   d) 90% ....

e) 7% ....

1.5	0.07	2.4	0.24	0.9	0.47	0.7	0.09	0.15

2 Draw a ring around the values in this list that are equivalent to $\dfrac{3}{20}$.

32%   3.2   0.15   0.03   23%   15%   0.23   3.2

3 Write these newspaper headlines using a percentage.

THREE OUT OF EVERY 10 CHILDREN DON'T VISIT THE DENTIST.

....

....

ONE FIFTH OF ADULTS CANNOT SWIM.

....

....

4 Write each decimal as a fraction in its simplest form.

a) 0.26 ....

b) 0.24 ....

c) 0.02 ....

d) 0.71 ....

e) 0.65 ....

f) 0.4 ....

5 Choose from the box an equivalent fraction, decimal or percentage for each value.

a) $\frac{7}{25}$ = ...

b) 18% = ...

c) $\frac{47}{50}$ = ...

d) 35% = ...

$\frac{7}{20}$	0.02	55%	0.7
25%	$\frac{9}{25}$	0.18	28%
0.72	$\frac{18}{25}$	94%	

6 Match each fraction with the equivalent percentage.

$\frac{3}{5}$		76%
$\frac{7}{100}$		65%
$\frac{3}{4}$		60%
$\frac{3}{50}$		75%
$\frac{13}{20}$		7%
$\frac{19}{25}$		6%

7 Use the digits 0, 1, 2, 3, 4, 5 and 8 once only to complete these two statements.

$$\frac{\boxed{}}{\boxed{}\ \boxed{}} = 0.\boxed{}\,5$$

$$\frac{\boxed{}}{\boxed{}} = \boxed{}\,0\%$$

16.2 Equivalent fractions, decimals and percentages 2

Summary of key points

Percentages can be greater than 100%.

$1.3 = 130\%$ $\qquad\qquad 4\frac{1}{5} = 420\%$ $\qquad\qquad 1.04 = 104\%$

Percentages can have decimal places.

$\frac{9}{1000} = 0.9\%$ $\qquad\qquad 0.013 = 1.3\%$ $\qquad\qquad 1.805 = 180.5\%$

Some fractions can be written with a denominator of 1000. You can convert these to terminating decimals or percentages.

$$\frac{101}{250} = \frac{404}{1000} = 0.404 = 40.4\%$$

Sometimes you cannot make the denominator 10, 100 or 1000.
To change any fraction to a decimal, divide the numerator by the denominator.

$$\frac{5}{16} = 5 \div 16$$

$$\begin{array}{r} 0.\ 3\ 1\ \ 2\ \ 5 \\ 16\ \overline{\smash{\big)}\ 5.^50\ ^20\ ^40\ ^80} \end{array}$$

$$\frac{5}{16} = 0.3125$$

Exercise 2

1 Write each decimal as a percentage.

a) 3.3 ....

b) 0.303 ....

c) 3.033 ....

d) 0.003 ....

2 Write each decimal or percentage as a fraction, simplifying where possible. Write fractions greater than 1 as mixed numbers.

a) 10.2 ....

b) 1.16 ....

c) 0.305 ....

d) 5.019 ....

e) 4.5% ....

f) 160% ....

g) 212.5% ....

h) 0.4% ....

3 Use a written method to convert these fractions to decimals.

a) $\frac{5}{8}$

b) $\frac{1}{16}$

c) $\frac{17}{40}$

....

4 For his homework, Paulo did the conversions below.

Mark each answer as correct or incorrect.

$\frac{3}{8} = 37.5\%$ $4.2 = 4.2\%$ $185\% = 1.85$ $2\frac{1}{4} = 2.4\%$

5 Find a decimal that is between $\frac{9}{20}$ and $\frac{113}{250}$. Show how you worked out your answer.

....

6 Use the digits 3, 6, 7 and 9 exactly once each to complete this number statement.

$$\frac{\boxed{}\boxed{}}{125} = 2\boxed{}.\boxed{}\%$$

16.3 Comparing decimals and fractions

Summary of key points

Using inequality signs

< means 'less than'	> means 'greater than'
= means 'equal to'	≠ means 'not equal to'

Comparing decimals

To compare the size of two or more decimals, first compare the digits with the largest place value.

0.38 > 0.291 because 3 > 2

If these two digits are the same, compare the digits with the second largest place value, and so on.

0.503 > 0.501 because 3 > 1

Comparing fractions

The size of two or more fractions can be compared by:

- writing them with a common denominator, or
- changing them to decimals.

Example:

Write the fractions $\frac{11}{16}, \frac{53}{80}$ and $\frac{13}{20}$ in order of size, smallest first.

The denominators (16, 80 and 20) are all factors of 80.

$\frac{11}{16} \xrightarrow[\text{multiply by 5}]{\text{multiply by 5}} \frac{55}{80}$ $\frac{53}{80} \xrightarrow[\text{no change}]{\text{no change}} \frac{53}{80}$ $\frac{13}{20} \xrightarrow[\text{multiply by 4}]{\text{multiply by 4}} \frac{52}{80}$

This shows that the correct order is $\frac{13}{20}, \frac{53}{80}, \frac{11}{16}$.

Exercise 3

1 Write an inequality sign (< or >) to make each statement true.

a) 0.22 .... 0.25   b) 0.012 .... 0.02   c) 3.036 .... 3.02

d) 1.005 .... 1.0046   e) 0.7498 .... 0.749 52   f) 0.0405 .... 0.041

2 Write each set of numbers in order <u>from largest to smallest</u>.

a) 0.14   0.138   0.11   0.097   ....

b) 0.024   0.042   0.005   0.03   ....

c) 1.998   1.98   1.99   1.989   ....

3 Write = or ≠ to make each statement correct.

a) $2\frac{1}{4}$ .... $\frac{6}{4}$ b) $\frac{22}{16}$ .... $1\frac{3}{8}$ c) $2\frac{5}{12}$ .... $\frac{58}{24}$ d) $1\frac{2}{17}$ .... $1\frac{3}{18}$

4 Write each set of fractions in increasing order.

a) $\frac{1}{4}$ $\frac{1}{8}$ $\frac{5}{32}$ $\frac{3}{16}$ ....

b) $\frac{117}{100}$ $\frac{5}{4}$ $1\frac{4}{25}$ $1\frac{9}{50}$ ....

c) Describe the methods you used for answering parts a) and b). Explain whether these were the most efficient methods.

....

....

....

5 Bella answered the following question:

Write = or ≠ to complete the statement: $\frac{5}{16}$ .... $\frac{3}{8}$

Bella's working is shown below.

$$0.3125$$
$$16 \overline{\smash{\big)}\ 5.\,^5 0\ ^2 0\ ^4 0\ ^8 0}$$

$$0.375$$
$$8 \overline{\smash{\big)}\ 3.\,^3 0\ ^6 0\ ^4 0}$$

$0.3125 \neq 0.375$, so $\frac{5}{16} \neq \frac{3}{8}$

Is Bella correct? Write a more efficient method for answering the question.

Correct ☐   Incorrect ☐

6 Zara says there is only one decimal number (n) for which $1.36 < n < 1.38$.

Is Zara correct? Explain your answer.

....

....

7 Write a whole number in each box to make the statement true.

$$\frac{1}{2} < \frac{\boxed{}}{8} = \frac{7}{\boxed{}} < \frac{\boxed{}}{20} < \frac{2}{3}$$

Think about

8 Write some number statements involving decimals and fractions to show your understanding of each of these signs: <, >, =, ≠.

17 Probability 1

You will practice how to:

- Use the language associated with probability and proportion to describe, compare, order and interpret the likelihood of outcomes.
- Understand and explain that probabilities range from 0 to 1, and can be represented as proper fractions, decimals and percentages.
- Identify all the possible mutually exclusive outcomes of a single event, and recognise when they are equally likely to happen.
- Understand how to find the theoretical probabilities of equally likely outcomes.

17.1 The language of probability

Summary of key points

- An **outcome** is the result of an **event**.
- For example, the outcomes from tossing a coin are a head or a tail. The event is tossing the coin.
- A **probability** is the chance that a particular outcome will happen.
- **Fair** means that all outcomes are equally likely.
- Probabilities can be compared on a **probability scale**. This is a scale from 0 to 1. An outcome that is impossible has probability 0. An outcome that is certain to happen has probability 1.

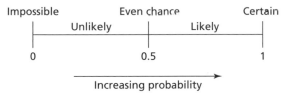

- Probabilities can never be greater than 1 or less than 0.
- Probabilities should not be written as ratios, such as 1 : 5, or in words, such as 1 in 3.

Exercise 1

1 **a)** **Put the following outcomes in order of probability from the least likely to most likely.**

    A   When a coin is thrown it lands on 'heads'.

    B   A person lives to be 1000 years old.

    C   You get homework sometime next week.

    D   It gets dark at some time tonight.

    E   A person chosen randomly is left-handed.

    ....

    least likely   most likely

b) Use one of these probability words to describe the likelihood of each outcome happening.

Impossible   Unlikely   Even chance   Likely   Certain

A .... ,   B ....   C ....   D .... E ....

2 A fair dice is thrown. Put these outcomes in order of probability from least likely to most likely.

A The dice shows 4.

B The dice shows an odd number.

C The dice shows a number greater than 2.

D The dice shows 8.

....

least likely   most likely

Use a probability word to describe the likelihood of each outcome happening.

3 Shade each spinner so that it matches the description.

a)   **b)**

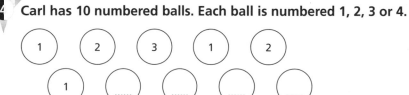

A ....   B ....

There is an even chance that the spinner lands on a red section.

It is unlikely that the spinner will land on a green section.

4 Carl has 10 numbered balls. Each ball is numbered 1, 2, 3 or 4.

(1)   (2)   (3)   (1)   (2)

(1)   (......)   (......)   (......)   (......)

When Carl chooses a ball at random:

- there is an even chance it will be a 1
- he has the same chance of picking a 3 and of picking a 4.

Number the rest of the balls.

5 **Here is a probability scale with four probabilities marked with arrows.**

Write which arrow points to the probability that:

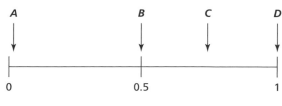

a) the next baby born is a boy ....

b) next Thursday it will rain all day without

stopping ....

c) a supermarket will sell some bread today ....

d) when you flip two coins, at least one will show heads ....

6 **A dice is rolled.**

Decide if each statement is true or false. Give a reason for each answer.

a) It is probable that the number rolled will be greater than 4.

....

....

b) A prime number outcome and an even number outcome are equally likely.

....

....

c) It is possible that the number rolled will be a multiple of 6.

....

....

7 **Anushka has this spinner.**

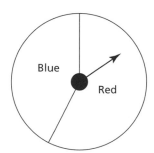

a) Mark on the scale with an arrow the probability of spinning Red.

0   0.5   1

b) Give a reason why Anushka's spinner is not fair.

....

....

8 Nazaneen picks one of these six cards at random.

Mark the probability of these outcomes on the scale.

A: The card is marked with an arrow.

B: The card is marked with an arrow pointing down (↓)

C: The card is marked with an arrow pointing up (↑)

D: The card is marked with an arrow pointing left (←)

9 Paulo has eight coloured balls in a bag. Each ball is either red or blue or green.

The probability scale shows the probability he picks out a ball of each colour.

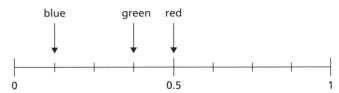

Write how many balls he has of each colour.

Red .... Blue .... Green ....

Think about

10 Make up your own probability outcomes to match each of these probability words:

Certain, Likely, Probable, Possible, Uncertain.

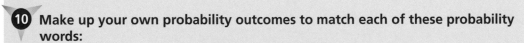

17.2 Listing outcomes

Summary of key points

The possible **outcomes** of an **event** are the possible results when the event takes place. For example, there are 6 possible outcomes from rolling a six-sided dice. These are 1, 2, 3, 4, 5 or 6.

Two outcomes are **mutually exclusive** if they cannot happen at the same time. For example, you cannot roll a 4 and a 6 on a single roll of a six-sided dice, so rolling a 4 and rolling a 6 are mutually exclusive.

1 A letter is chosen at random from each of the following words. List the possible outcomes. In each case, say whether the outcomes are equally likely or not.

a) TODAY ....

b) TOMORROW ....

2 Tick the pairs of outcomes that cannot happen at the same time.

Getting a 1 when a dice is thrown	Getting an odd number when a dice is thrown	☐
Getting an even number when a dice is thrown	Getting an odd number when a dice is thrown	☐
Winning a game of chess	Losing a game of chess	☐
Having pasta for dinner tomorrow	Having salad for dinner tomorrow	☐

3 Gareth picks one of these cards at random.

a) List the possible outcomes for the number of sides on the shape he picks.

....

b) List the possible outcomes for the colour of the shape he picks.

....

c) Gareth says that getting a white shape and getting a shape with 3 sides are mutually exclusive.

Is he correct? Give a reason for your answer ....

....

....

4 A bag contains the numbers 1–10.

A number is picked at random.

For each pair of events, decide if between them they cover all the possible outcomes.

		Cover all possible outcomes	Do not cover all possible outcomes
Getting an odd number	Getting an even number	☐	☐
Getting a number less than 5	Getting a number more than 5	☐	☐
Getting a square number	Getting a multiple of 3	☐	☐

5 Tick to show if the outcomes in each situation are equally likely or not.

	Equally likely	Not equally likely
Getting a head or tail when a coin is thrown.	☐	☐
Getting a win, draw or loss in a football match.	☐	☐
Getting a 1, 2, 3, 4, 5 or 6 when a dice is thrown.	☐	☐

Think about

6 A letter is picked from a word at random. There are 5 outcomes. What could the word be if the 5 outcomes are equally likely? What could the word be if the 5 outcomes are not equally likely?

17.3 Calculating probabilities

Summary of key points

When all outcomes of an event are equally likely we can use this formula:

probability of a particular outcome = $\dfrac{\text{number of favourable outcomes}}{\text{total number of outcomes}}$.

Probabilities can be given as fractions or decimals.

Jim has these cards.

| S | Q | U | A | R | E | S |

He picks a card at random.

a) How many possible outcomes are there? ....

b) Write the probability that he picks:

i) the letter R ....

ii) the letter S ....

iii) the letter T ....

Ellie has 9 balls. Each ball is labelled with a letter.

A U S T R A L I A

She picks a ball at random. Write down the probability that it is labelled with:

a) the letter L ....   b) the letter U ....

c) the letter A ....   d) the letter Z ....

3 Mika has these cards.

| ▲ | ▲ | ▲ | ▲ | ■ | ■ | ● |

He chooses a card at random.

a) Write the probability that the card is marked with:

i) a circle ....   ii) a square ....

iii) a triangle ....   iv) a pentagon ....

Mika uses the cards to play a game with Jacob.

They take it in turns to randomly select a card. Mika wins if he selects a card with a triangle. Jacob wins if he selects a card with a square or a circle.

b) Is the game fair? Explain your answer.

4 Jana has a jar containing 10 coloured beads.

3 are green   6 are blue   1 is red

She picks a bead at random. Write as a decimal the probability that the bead is:

a) red   ....

b) blue   ....

c) yellow   ....

d) green or blue   ....

5 Hazim spins the spinner.

Write the probability that the spinner points to:

a) the number 8 ....

b) the number 5 ....

c) a number less than 3 ....

d) a number more than 5 ....

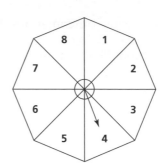

6 A jar contains 10 balloons.

The probability of choosing each colour of balloon is shown in the table.

Colour	Probability
Green	$\frac{1}{2}$
Blue	$\frac{1}{10}$
Red	$\frac{2}{5}$

Work out how many balloons of each colour are in the jar.

Green ....   Blue ....   Red ....

7 The table shows the colours of cars in a car park.

Colour	Number of cars
Black	12
Red	14
Blue	13
Other	11

One car is chosen at random. Work out, as a decimal, the probability that it is:

a) red ....

b) blue ....

c) black or red ....

d) neither black nor red nor blue. ....

8 Zach picks one of these cards at random.

Draw a line to match each event with the correct probability.

Zach picks a card with a circle.

$\dfrac{1}{2}$

Zach picks a card with a rectangle and the number 3.

$\dfrac{3}{4}$

Zach picks a card with the number 3.

$\dfrac{1}{4}$

Zach picks a card with a shape that has 3 or 4 sides.

$\dfrac{1}{6}$

9 There are 20 cards in a pack.

All the cards are coloured either red or blue or green.

6 cards are coloured red. There are equal numbers of cards coloured blue and green.

A card is chosen at random.

Find the probability that the card is:

a) red .... **b)** green ....

10 Pria is given a bunch of flowers containing only roses and carnations. The table shows the number of stems of each type of flower.

	Red	Yellow	White
Roses	4	1	3
Carnations	2	2	4

Pria chooses one stem from the bunch at random.

Are these statements true or false?

	True	False
a) The stem is equally likely to be a rose or a carnation.	☐	☐
b) The stem is more likely to be red than white.	☐	☐
c) The probablility of picking a red rose is $\frac{1}{4}$.	☐	☐
d) The probablility of picking a yellow carnation is $\frac{9}{16}$.	☐	☐

11 The table shows some information about Sophie's books.

	Type of books		
	Fiction	Historical	Nature
Paperback	40	20	4
Hardback	20	8	8

A book is selected at random. Complete these sentences.

a) The probability that the book is a historical paperback is ....

b) The probability that the book is a .... book is 0.4.

c) The probability that the book is a .... is 0.28.

d) The probability that the book is a paperback and a .... book is 0.04.

Think about

12 Design a spinner. Colour some of the sections in red, some in blue and the rest in green. Write the probability that the spinner lands on each colour.

Is your spinner biased? Explain your answer.

Transformations

You will practice how to:

- Use knowledge of 2D shapes and coordinates to find the distance between two coordinates that have the same *x* or *y* coordinate (without the aid of a grid).
- Use knowledge of translation of 2D shapes to identify the corresponding points between the original and the translated image, without the use of a grid.
- Reflect 2D shapes on coordinate grids, in a given mirror line (*x*- or *y*-axis), recognising that the image is congruent to the object after a reflection.
- Rotate shapes 90° and 180° around a centre of rotation, recognising that the image is congruent to the object after a rotation.
- Understand that the image is mathematically similar to the object after enlargement. Use positive integer scale factors to perform and identify enlargements.

18.1 Translations

Summary of key points

The distance between two coordinates with the same *x*- or *y*-coordinates is found by finding the difference between the other two coordinates.

For example,   *A* (3, 4) and *B* (3, 8) both lie on a vertical line. The distance between them is 8 – 4 = 4 units.

C (1, 7) and *D* (5, 7) both lie on a horizontal line. The distance between them is 5 – 1 = 4 units.

A **translation** is a movement of a shape.

You usually state the size and direction of the movement horizontally (left/right) and vertically (up/down).

You call the shape you start with the **original** shape or the **object**. You call the translated shape the **image**.

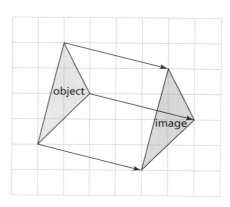

Exercise 1

1 **Find the distances between the following coordinates.**

    a) (2, 8) and (7, 8) ....

    b) (–3, 5) and (4, 5) ....

    c) (3, 9) and (3, 1) ....

    d) (–1, –2) and (–1, –5) ....

2 The distance between two points A and B is 4 units. Tick the correct box for each statement.

	True	False
a) The coordinates of A and B could be (3, 8) and (3, 7)	☐	☐
b) The coordinates of A and B could be (–2, 6) and (2, 6)	☐	☐
c) The coordinates of A and B could be (8, 1) and (8, –3)	☐	☐

3 Draw a line from each pair of points to the appropriate description.

(4, 6) and (4, 11)

(–1, –3) and (–1, –1)

Length between is less than 5 units

(2, 6) and (–1, 6)

(1, –2) and (8, –2)

Length between is equal to 5 units

(–1, –3) and (–1, 8)

Length between is greater than 5 units

(–5, 4) and (–10, 4)

4 The coordinates of the vertices of a triangle ABC are A(3, 1), B(–4, 5), C(2, –1).

The triangle is translated 4 units left and 3 units down.

Write down the vertices of the image.

image of A(3, 1) is ....

image of B(–4, 5) is ....

image of C(2, –1) is ....

5 The image of the point (–2, 4) under a translation is (6, –1).

Find the coordinates of the image of (1, 5) under the same translation.

....

6 ***ABCD* is a parallelogram. The coordinates of *A* and *B* are *A*(3, 5) and *B*(7, 8).**

The translation from *B* to *C* is 2 right and 3 up.

   a) Write down the coordinates of *C*. ....

   b) Write down the coordinates of *D*. ....

7 ***ABC* is an isosceles triangle. *A* has coordinates (2,5). *B* has coordinates (6,9).**

Write down two possible coordinates for point *C*.

.... or ....

18.2 Reflections and rotations

Summary of key points

A **reflection** is a mirroring or flipping of a shape.

There is always a **mirror line** for a reflection.

A **rotation** is the turning of a shape.

You should give the angle and direction of the rotation as well as the centre point that the shape turns around.

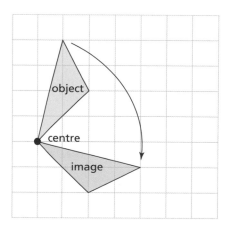

When a shape is transformed using a translation, reflection or rotation, the image will be congruent to the object.

1. **a)** Reflect shape A in the *y*-axis.
   Label the image B.

   b) Reflect shape A in the *x*-axis.
   Label the image C.

   c) Reflect shape B in the *x*-axis.
   Label the image D.

   d) Write down the shapes that are congruent to A.

   ....

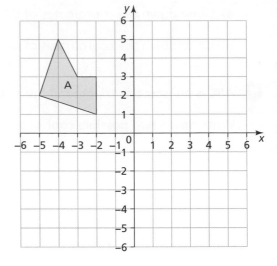

2. **a)** Reflect shape P in the *y*-axis and label it Q.

   b) Reflect shape P in the *x*-axis and label it R.

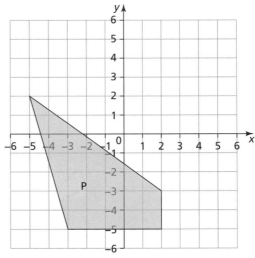

3. **The point *B* is the image of point *A* after a reflection in the *x*-axis. Tick to show if these statements are true or false.**

	True	False
a) The line *AB* is perpendicular to the *x*-axis.	☐	☐
b) *B* is the same distance from the *x*-axis as *A*.	☐	☐
c) *A* is the same distance from the *y*-axis as *B*.	☐	☐
d) If *A* is (15, 21) then *B* will be (–15, 21).	☐	☐

4 Rotate the shaded shape using the centre and angle given.

a) (Rotate by 180° centre point *P*)

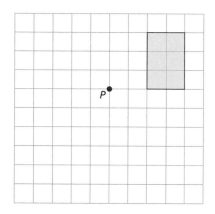

b) (Rotate by 90° clockwise centre point *P*)

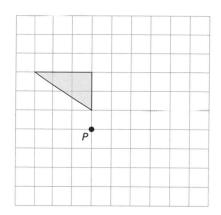

c) (Rotate by 90° anticlockwise centre point *P*)

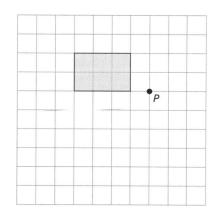

d) (Rotate by 90° anticlockwise centre point *P*)

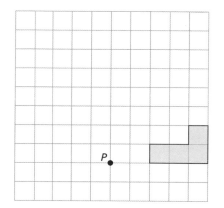

e) (Rotate by 180° centre point *P*)

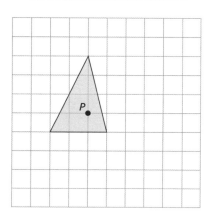

f) (Rotate by 90° clockwise centre point *P*)

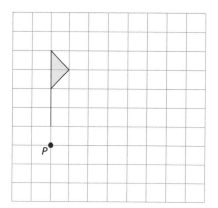

5 The diagram shows two triangles **T** and **U**.

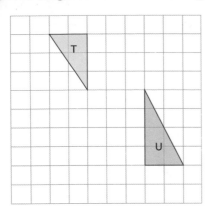

Pat says that **U** is a rotation of **T**. Explain why Pat must be wrong.

....

....

Think about

6 Draw a picture on a square grid. Transform your picture using a reflection, a rotation and a translation.

18.3 Enlargements

Summary of key points

An **enlargement** transforms a shape by making it larger (or smaller).

The **scale factor** of an enlargement tells us how many times larger the image is.

The image will be **similar** to the original shape. This means that the object and the image will have the same shape.

Translations, reflections, rotations and enlargements are all types of **transformations**.

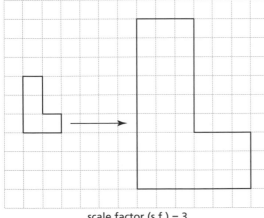

scale factor (s.f.) = 3

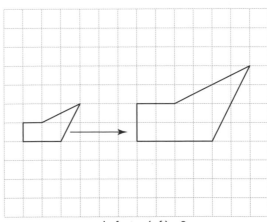

scale factor (s.f.) = 2

1 Colour the shapes that are enlargements of shape S.

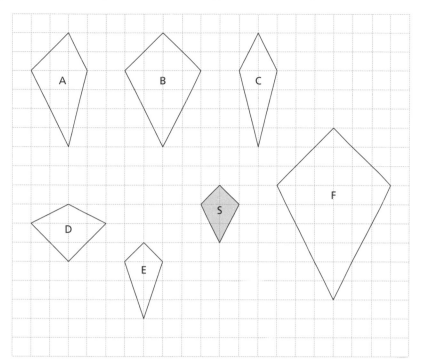

2 Write down the letters of the shapes that are enlargements of triangle T.

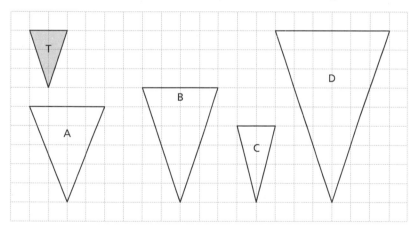

....

3 Draw the enlargements of each shape using the scale factor given. One side of the enlargement has been drawn for you.

a)

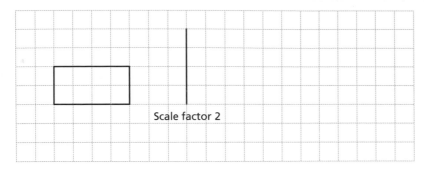

Scale factor 2

b)

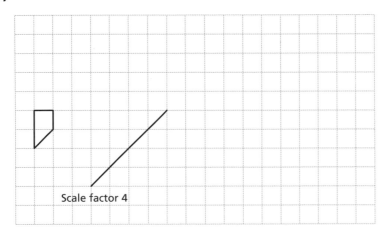

Scale factor 4

c)

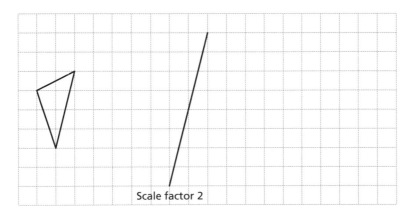

Scale factor 2

4 Explain how you can tell that rectangle S is not an enlargement of rectangle R.

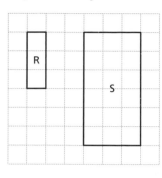

....

....

5 *A'B'C'D'* is an enlargement of quadrilateral *ABCD*. The diagram shows some of the points.

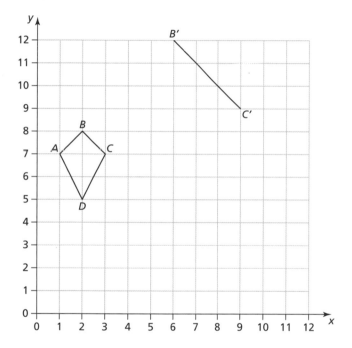

a) Complete *A'B'C'D'*.

b) Write down the scale factor of the enlargement. ....

6 Are these statements about enlargements true or false?

	True	False
a) The angles in the image are each larger than the corresponding angles in the object.	☐	☐
b) When you enlarge a regular pentagon, the image is always another regular pentagon.	☐	☐
c) When you enlarge a shape, the object and the image are mathematically similar.	☐	☐

Percentages

You will practice how to:

- Recognise percentages of shapes and whole numbers, including percentages less than 1 or greater than 100.

19.1 Percentages of quantities

Summary of key points

Below are examples of working out percentages of a quantity without a calculator.

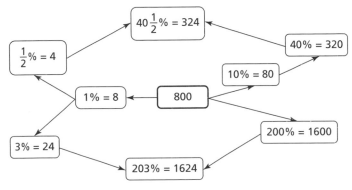

To find a percentage of a quantity with a calculator, multiply the quantity by the percentage (using the percentage key), or by the decimal or fraction equivalent of the percentage.

Exercise 1

1–6, 8

1 Find:

    a) 6% of 300   **b)** 12% of $4000   **c)** 7% of 1200 kg   **d)** 34% of 600

....

2 Put a ring around the calculation in each set that gives a different answer to the other two.

    a) 25% of 128   1.5% of 2000   12.5% of 240

    b) 24% of 600   95% of 160   150% of 96

3 Find:

   a) 175% of 48 ....   **b)** 3.5% of 60 ....

4 25% of a number is 18. What is the number? ....

5 Use the digits 1, 2, 3, 5 and 6 to complete this statement.

   ☐☐ % of 360 = ☐☐☐

6 Jade is asked to find 18.5% of 420. Her working is shown below.

20% of 420 = 420 ÷ 20 = 21

1% of 420 = 420 ÷ 100 = 4.2
0.5% of 420 = 2.1
So 1.5% of 420 = 4.2 + 2.1 = 6.3
So 18.5% of 420 = 21 − 6.3 = 21 − 6 + 0.3 = 15.3

What mistakes has Jade made?

....

....

....

....

7 Use a calculator to find

   a) 34.2% of $21 500 ....

   b) 0.3% of 3800 metres ....

   c) 285% of $660 ....

   d) 0.75% of 4800 ....

Think about

8 Write down 5 percentage calculations that give an answer of 24.

One of your calculations should involve a percentage less than 1% and one should involve a percentage greater than 100%.

Summary of key points

To write one quantity as a percentage of a second quantity, e.g. 6 as a percentage of 30...

without a calculator

using a calculator

write the first quantity as a fraction of the second: $\frac{6}{30}$

divide the first number by the second number: $6 \div 30 = 0.2$

express as an equivalent fraction with denominator 100: $\frac{6}{30} = \frac{1}{5} = \frac{20}{100}$

multiply by 100 to find the equivalent percentage:

$0.2 \times 100 = 20$, so 20%

write as the equivalent percentage: 20%

Exercise 2

1–4

1. **Find the percentage of each shape that has been shaded.**

   a)

   ....%

   b)

   ....%

   c)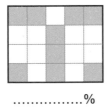

   ....%

   d)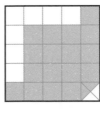

   ....%

2. **Shade 40% of each shape.**

   a)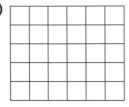

   b)

3 Write these as percentages.

a) $42 out of $100 ....%

b) 8 grams out of 50 grams ....%

c) 130 marks out of 200 marks ....%

d) 7 days out of 10 days ....%

e) 14 children out of 25 children ....%

4 Complete the table, writing the first number as a percentage of the second.

First number	Second number	Percentage
0.5	100	
1	1000	
66	55	
88	20	

5 Khaled has 80 books.

a) 58 of the books are paperback books. What percentage of all the books are paperbacks?

.........%

b) 16 of the books are history books. What percentage of all the books are history books?

.........%

c) 14 of the history books are paperbacks. What percentage of the history books are paperbacks?

.........%

6 Put a ring around the proportion that is different to the other two.

a) 70%   72 out of 96   98 out of 140

b) 39 out of 60   65%   330 out of 550

7 Complete these statements.

.........% of 250 = 150   25% of 120 = ...% of 200

Presenting and interpreting data 2

You will practice how to:

- Record, organise and represent categorical, discrete and continuous data. Choose and explain which representation to use in a given situation:
  - o waffle diagrams and pie charts
  - o line graphs
  - o scatter graphs
  - o infographics.
- Interpret data, identifying patterns, within and between data sets, to answer statistical questions. Discuss conclusions, considering the sources of variation, including sampling, and check predictions.

20.1 Waffle diagrams and pie charts

Summary of key points

Waffle diagrams and **pie charts** are used to visualise how a population is made up. They are diagrams that are used to show proportions.

Waffle diagrams are coloured grids.

In a pie chart, a circle is used to represent the whole data. It is divided into sectors. Each sector needs to be labelled.

Exercise 1

1 **Some people were asked what type of television programme they most liked to watch.**

A pie chart showing the most popular types of television programmes

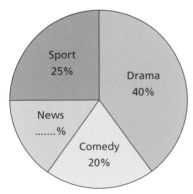

a) Which type of programme was twice as popular as comedy? ....

b) What percentage of people chose either drama or sport? ....%

c) What percentage of people chose news? ....%

2 There are 16 boys and 16 girls in a class.

At breaktime, all the children chose a drink.

The pie charts show their choices.

Find the total number of children who chose milk.

....

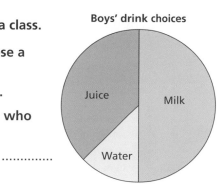

Boys' drink choices

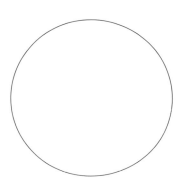

Girls' drink choices

3 The table shows the favourite flavours of ice cream for 20 children.

Show the information as a pie chart.

Flavour	Frequency
Vanilla	7
Strawberry	6
Chocolate	4
Mint	3

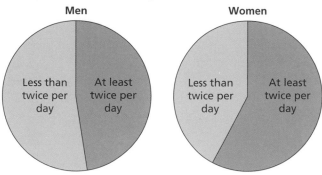

4 Harry is investigating how often people clean their teeth. He asks 40 men and 40 women whether they clean their teeth at least twice a day.

Men

Less than twice per day | At least twice per day

Women

Less than twice per day | At least twice per day

Write some conclusions based on Harry's pie charts.

....

....

....

5 The waffle diagram shows the types of pizzas ordered by restaurant customers one day.

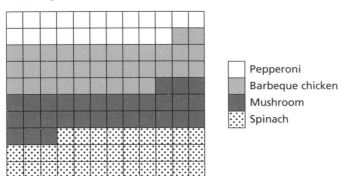

Pepperoni
Barbeque chicken
Mushroom
Spinach

Each square represents two pizzas.

a) Find how many customers chose a pepperoni pizza.

....

b) Write down the type of pizza that had the most orders.

....

c) Show that 25% of the pizzas ordered were mushroom.

....

....

....

6 The pie chart shows the sizes of coats sold by a shop during one week.

The shop sold 18 small coats.

How many large coats did the shop sell in this week?

....

Coat sales during one week

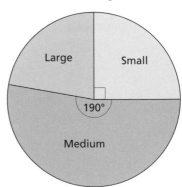

7 A shop sells three types of milk. The table shows the number of litres of each type of milk it sells one day.

Type of milk	Number of litres sold
Full fat	112
Reduced fat	264
Very low fat	144

Draw a waffle diagram to show the information. Use one square to represent four litres.

Remember to give a key.

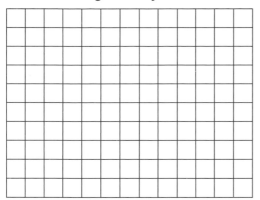

Think about

3 Use a spreadsheet to draw a pie chart to show how many hours you spent on different activities each day. Compare your pie chart with a partner's work.

20.2 Infographics

Summary of key points

An **infographic** is a general term for a diagram that is designed to display data in a visually appealing way. An infographic should display the information in a very clear way that makes interpretation quick and easy.

Exercise 2

Real data question. The infographic shows U.S. greenhouse gas emissions in 2017.

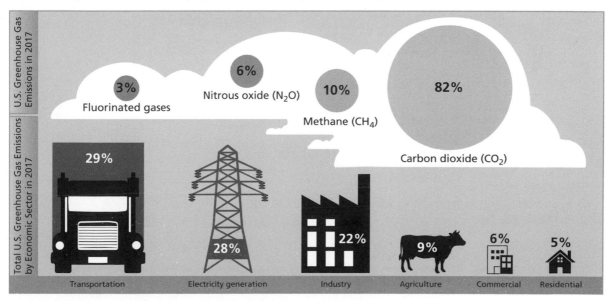

Source: United States Environmental Protection Agency

State whether the following conclusions are true, false or if you cannot tell from the infographic.

Give a reason for each answer.

a) Cars produced almost a quarter of greenhouse gases. True ☐ False ☐ Cannot tell ☐

....

b) Electricity generation and transport combined produced more than half of greenhouse gas emissions. True ☐ False ☐ Cannot tell ☐

....

c) Agriculture and commercial together produced $\frac{3}{20}$ of all emissions. True ☐ False ☐ Cannot tell ☐

d) Which greenhouse gas is produced the most?

....

2 **Real data question. The infographic shows the altitudes and distance covered in 'La Marmotte', an annual cycling race in France. Distances are shown in kilometres and altitude in metres.**

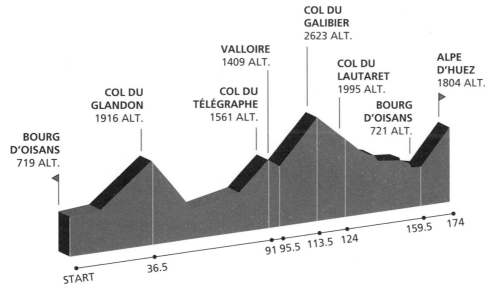

Source: Marmotte Granfondo Series

a) Write down the highest altitude reached during the race.

.... metres

b) Find how far it is from the top of the Col du Glandon to the top of the Col du Télégraphe.

.... kilometres

c) Find the gain in altitude from Valloire to the top of the Col du Galibier.

.... metres

3 Real data question. The pie charts show the type of fuel used by cars in 1994 and 2016.

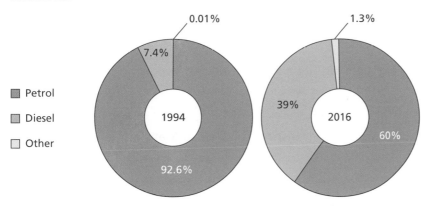

0.01%

7.4%

1.3%

39%

2016

60%

Petrol

Diesel

Other

1994

92.6%

Source: © Crown copyright 2017

State whether the following conclusions are true, false or if you cannot tell from the infographic.

Give a reason for each answer.

a) In 2016, more cars used diesel fuel than petrol.   True ☐   False ☐   Cannot tell ☐

....

b) There were more cars on the roads in 2016 than 1994.   True ☐   False ☐   Cannot tell ☐

....

c) $\frac{3}{5}$ of all cars on the road in 2016 used petrol.   True ☐   False ☐   Cannot tell ☐

....

d) There were 7400 diesel vehicles on roads in 1994.   True ☐   False ☐   Cannot tell ☐

....

4 Real data question. The infographic shows the different modes of transport used by passengers to travel to airports in the UK.

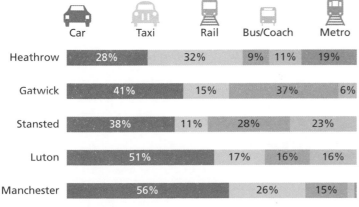

	Car	Taxi	Rail	Bus/Coach	Metro
Heathrow	28%	32%	9%	11%	19%
Gatwick	41%	15%	37%	6%	
Stansted	38%	11%	28%	23%	
Luton	51%	17%	16%	16%	
Manchester	56%	26%	15%		

Source: © Crown copyright 2017

a) Write down which mode of transport was used most to travel to Heathrow.

....

b) Write down the airports that had over half of passengers arriving by car.

....

c) Find what percentage of passengers did not arrive at Gatwick airport by car or taxi.

....

d) Write down the least popular method of transport to arrive at Stansted.

....

20.3 Line graphs

Summary of key points

Line graphs are used to show changes in a quantity over time. Two or more line graphs can be plotted on the same axes to make comparisons.

Exercise 3

1 **Pam is investigating this question:**

Has the percentage of children that pass a maths examination increased every year since 2014?

She draws this graph to illustrate her data.

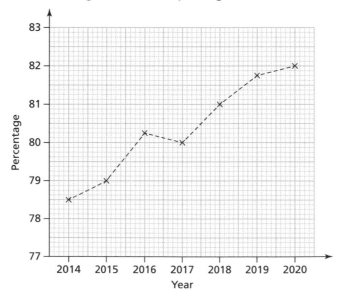

Percentage of students passing a maths exam

What conclusions can Pam draw from her data?

....

....

2 The line graph shows the mean monthly temperature in Brisbane.

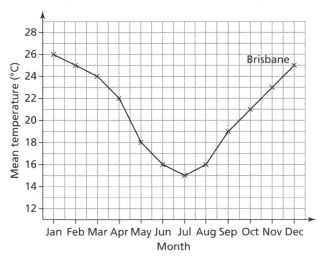

The table shows the mean monthly temperature in Malaga.

Month	Jan	Feb	Mar	Apr	May	Jun	Jul	Aug	Sep	Oct	Nov	Dec
Mean temperature (°C)	13	13	15	16	19	22	25	25	24	20	16	13

a) Add a line to the graph to show the mean monthly temperature in Malaga.

b) Write down the difference between the mean temperatures in Malaga and Brisbane in November.

.....................°C

c) Which is the first month in the year when the mean temperature in Malaga is greater than the mean temperature in Brisbane?

....

3 The mean daily temperatures in Dubai and Kiev are shown in the table.

City	Temperature (°C)											
	Jan	Feb	Mar	Apr	May	Jun	Jul	Aug	Sep	Oct	Nov	Dec
Dubai	19	20	23	26	31	34	35	35	33	30	25	21
Kiev	−3	−2	3	9	16	19	21	20	15	9	2	−2

a) Draw line graphs on the grid below to display the information for each city.

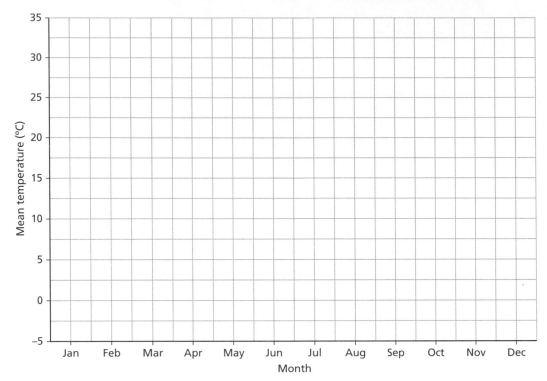

b) Find the range of temperatures in Dubai and Kiev over the course of a year.

Dubai .... °C   Kiev .... °C

c) How much colder is it in Kiev than Dubai in January?

....

4 **The temperature outside Claudia's house is recorded each hour.**

The line graph shows the measurements for one day.

a) What was the temperature at 13:00?
.........°C

b) Write down the highest temperature recorded. ....°C

c) Write down the times at which the recorded temperature was higher than 10°C. ....

d) At what time was the recorded temperature the same as it was at 01:00? ....

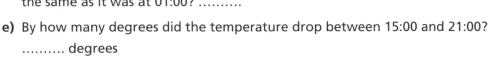

Temperature during one day

e) By how many degrees did the temperature drop between 15:00 and 21:00?
.... degrees

f) Describe what happened to the temperature during the first 8 hours of the day.

....

5 Norman is investigating this hypothesis:

The number of students at Manor High School increased more between 2013 and 2017 than the number of students at Valley College.

The information he found on the internet about pupil numbers at both schools is shown in the graph and in the table.

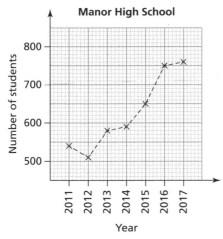

Valley College	2013	2015	2017
Number of boys	317	348	402
Number of girls	308	337	395

Is Norman's hypothesis true or false? True ☐ False ☐

Show how you worked out your answer.

....

....

....

Think about

6 Draw a line graph to show how the population of your town or country has changed over recent years.

Summary of key points

A **scatter graph** shows if two variables are related.

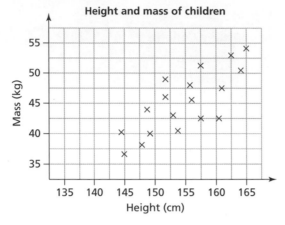

Height and mass of children

This scatter graph shows the heights and masses of 18 children. Taller children are generally heavier than shorter children.

Exercise 4

1 Tina records how long it takes 10 children to solve a puzzle. The scatter graph shows the times in seconds plotted against each child's age.

Are these statements true or false?

	True	False
a) Two children are 12 years old.		
b) The longest time taken to solve the puzzle was 65 seconds.		
c) The person who solved the puzzle quickest was 16 years old.		
d) Two children took 42 seconds to solve the puzzle.		
e) Older children tend to solve the puzzle quicker than younger children.		

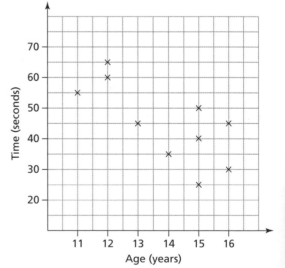

2 Some athletes took part in two sprints: one over 100 metres and one over 400 metres.

The table shows their times.

Time for 100 m (seconds)	12.2	13.3	12.5	13.6	14.5	13.7	14.2
Time for 400 m (seconds)	52.4	53.6	53.2	53.4	52.8	54.4	54.8

Draw a scatter graph to show these data.

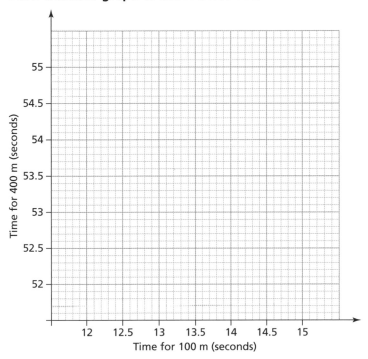

3 The scatter diagrams show some information about eight cars.

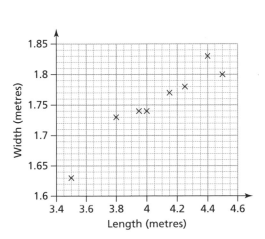
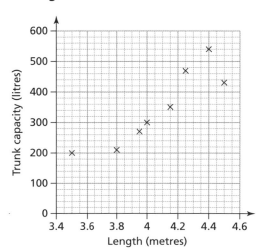

a) Write down the trunk capacity of the shortest car.

.... litres

b) One of the cars has a trunk capacity of 300 litres. Find the **width** of this car.

.... metres

4 The time in minutes that nine teenagers spent using their phone and spent studying on one day is shown in the table.

Time spent studying (minutes)	0	15	19	25	28	45	59	82	90
Time spent using phone (minutes)	80	65	60	45	43	32	21	16	10

a) Draw a scatter graph to show this information.

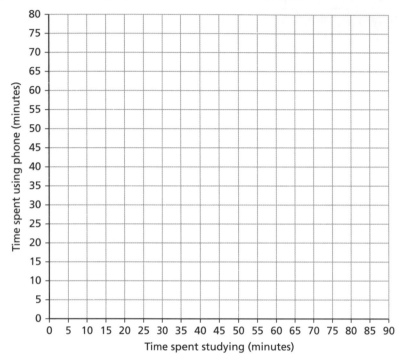

Time spent studying (minutes)

b) Draw a line of best fit on your graph.

c) Use your line of best fit to predict the time a teenager will spend studying if they spend 50 minutes per day on their phone.

.... minutes

d) Write down the relationship between time spent studying and time spent on the phone.

....

....

Equations and inequalities

You will practice how to:

- Understand that a situation can be represented either in words or as an equation. Move between the two representations and solve the equation (integer coefficients, unknown on one side).
- Understand that letters can represent an open interval (one term).

21.1 Forming and solving equations

Summary of key points

An **equation** is a mathematical statement that includes an equal sign (=). It is true for a particular value of the unknown quantity represented by the letter, and can be solved to find this value.

Example

$$4x - 2 = 26$$

$+2$ $+2$

$$4x = 28$$

$\div 4$ $\div 4$

$$x = 7$$

So the solution is $x = 7$.

Example

$$30 - 2x = 24$$

$+2x$ $+2x$

$$30 = 24 + 2x$$

-24 -24

$$6 = 2x$$

$\div 2$ $\div 2$

$$3 = x$$

The solution is $x = 3$.

Exercise 1

1 **Find the number that each symbol represents.**

a) $\blacktriangle \times 6 = 54$ $\blacktriangle = $ ....

b) $\bullet \div 3 = 10$ $\bullet = $ ....

c) $\blacksquare \times 2 + 1 = 19$ $\blacksquare = $ ....

d) $\lozenge \times 5 - 2 = 18$ $\lozenge = $ ....

2 **Solve:**

a) $x + 8 = 15$ $x = $ ....   b) $t - 6 = 11$ $t = $ ....

c) $4y = 44$ $y = $ ....   d) $6n = 30$ $n = $ ....

3 By solving each equation to find the value of x, place each equation in the correct column of the table. The first equation has been put in for you.

Equation A

$x + 2 = 11$

Equation B

$x - 6 = 12$

Equation C

$x + 4 = 10$

Equation D

$x - 8 = 4$

Equation E

$3x = 12$

Equation F

$6x = 36$

Equation G

$x \div 2 = 9$

Equation H

$6x = 54$

$x = 4$	$x = 6$	$x = 9$	$x = 12$	$x = 18$
		Equation A		

4 By solving each equation, find the odd equation out.

$7x = 35$ $9x = 36$ $x + 7 = 12$

5 Solve these equations.

a) $2x - 3 = 11$

$x = \ldots\ldots\ldots\ldots$

b) $4x + 5 = 13$

$x = \ldots\ldots\ldots\ldots$

c) $5x - 1 = 24$

$x = \ldots\ldots\ldots\ldots$

d) $2x + 8 = 16$

$x = \ldots\ldots\ldots\ldots$

e) $6x + 2 = 32$

$x = \ldots\ldots\ldots\ldots$

f) $5x + 3 = 43$

$x = \ldots\ldots\ldots\ldots$

6 Put a ring around the solution to the equation $20 - 2y = 12$.

$y = 4$ $y = 6$ $y = 10$ $y = 16$

7 Find the number hiding under each ink mark. Then find the value of y.

a)

$30 - \qquad = 24$

$\ldots\ldots\ldots\ldots$

$30 - 2y = 24$

$y = \ldots\ldots\ldots\ldots$

b)

$$18 - \text{✦} = 2$$

....

$$18 - 4y = 2$$

$$y = ...$$

8 Solve:

    a) $14 - 6t = 8$ **b)** $17 - 2w = 5$

                  $t = ...$ $w = ...$

9 Use ticks and crosses to mark Spencer's homework.

①	$5g = 20$	④ $20 - 4r = 12$
	$g = 100$	$4r = 32$
		$r = 8$
②	$2h + 3 = 17$	
	$2h = 14$	⑤ $4t + 5 = 25$
	$h = 7$	$4t = 20$
		$t = 5$
③	$3m - 6 = 18$	
	$3m = 12$	⑥ $16 - 5n = 6$
	$m = 4$	$5n = 10$
		$n = 2$

10

> Dave thinks of a number.
>
> He multiplies it by 4.
>
> He then adds 6.
>
> His answer is 34.

a) Put a ring around the equation that matches this number puzzle.

    $4(x + 6) = 34$ $4x + 6 = 34$ $6x + 4 = 34$ $6(x + 4) = 34$

b) Solve the equation to find the number Dave thought of.

....

11 By forming an equation, solve these number puzzles.

a)
> Talia thinks of a number.
>
> She multiplies it by 2.
>
> Then she adds 7.
>
> Her answer is 19.
>
> What number did Talia think of?

....

b)
> Sabah thinks of a number.
>
> She multiplies it by 5.
>
> Then she subtracts 3.
>
> Her answer is 42.
>
> What number did Sabah think of?

....

12 25 children are arranged into four teams.

Teams A, B and C each have the same number, n, of children.

Team D has 7 children.

a) Put a ring around the equation that describes this situation.

$4n + 7 = 25$ $4n - 7 = 25$ $3n + 7 = 25$ $3n - 7 = 25$

b) Solve the equation to find n.

....

13 Pencils can be bought in boxes or separately.

Marek buys three boxes of pencils. He also buys two loose pencils.

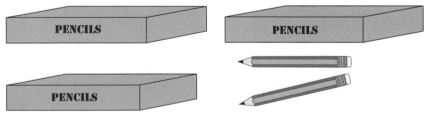

In total, he has 26 pencils.

By forming an equation, find how many pencils are in each box.

....

Make up five equations. The solution of each equation should be $x = 6$.

21.2 Inequalities

Summary of key points

The symbol '>' is used for '**greater than**' statements, and the symbol '<' is used for '**less than**' statements.

For example, $5 < 8$ (five is less than eight) or $m > 6$ (m is greater than six).

An inequality can be represented on a number line. For example, for $m < 8$ the number line would be:

```
0  1  2  3  4  5  6  7  8  9  10
```

The open circle at the end of the arrow shows that you go right up to 8, but you don't include it.

Exercise 2

1 Draw lines to match the inequality statements in words on the left and the algebra on the right.

f is less than 4	$f > 3$
g is more than 2	$g < 2$
f is more than 5	$f < 4$
g is less than 2	$g < 3$
f is more than 3	$f > 5$
g is less than 3	$g > 2$

2 Draw these inequalities on a number line.

a) $m < 5$

```
-6 -5 -4 -3 -2 -1  0  1  2  3  4  5  6
```

b) $t > 1$

```
-6 -5 -4 -3 -2 -1  0  1  2  3  4  5  6
```

c) $d > -2$

-6 -5 -4 -3 -2 -1 0 1 2 3 4 5 6

d) $k < 0$

-6 -5 -4 -3 -2 -1 0 1 2 3 4 5 6

3 Write the correct inequality in the gaps.

a) 25 30   **b)** 214 241   **c)** −3 ...... −6   **d)** $\dfrac{1}{2}$ ...... $\dfrac{2}{3}$

e) 0.64 ...... 0.637   **f)** 0.6 ...... $\dfrac{2}{3}$ **g)** 12.7 ...... 7.12   **h)** $-\dfrac{1}{2}$ .... −1

4 t is a number such that $t < 7$. Circle the possible values of t.

−12   10   0   6.99   7.01 $6\dfrac{1}{2}$

5 m is a positive whole number and $m < 20$.

m is also a multiple of 5.

List the possible values of m.

....

6 $y = 12.5$. Circle the inequalities that are true.

$y > 10$ $-30 > y$ $y > 12.6$ $15 > y$

$y < 12.6$ $y > -14$ $0 < y$

7 A giraffe weighs g kilograms. A rhinoceros weighs r kilograms and a hippopotamus weighs 1600 kg.

a) A rhinoceros weighs more than a hippopotamus. Write an inequality to show this.

....

b) A giraffe weighs less than a hippopotamus. Write an inequality to show this.

....

c) Which animal weighs the least?

....

8 Kumal thinks of an integer, *n*.

a) His number is more than –3. Write an inequality to show this.

....

b) His number is less than 3. Write an inequality to show this.

....

c) Draw both of these inequalities on a number line.

d) What are the possible values of Kumal's number?

....

9 *p* is a prime number.

a) *p* is more than 30. Write an inequality to show this.

....

b) *p* is less than 40. Write an inequality to show this.

....

c) List the possible values of *p*.

....

10 Matt is thinking of two inequalities about a number, *x*.

The only integers that fit both inequalities at the same time are **2 and 3**.

Write down the two inequalities that Matt is thinking of.

....

Show his inequalities on the number line.

Ratio and proportion

You will practice how to:

- Use knowledge of equivalence to simplify and compare ratios (same units).
- Understand how ratios are used to compare quantities to divide an amount into a given ratio with two parts.
- Understand and use the unitary method to solve problems involving ratio and proportion in a range of contexts.

22.1 Simplifying and comparing ratios

Summary of key points

In the **simplest form** of a ratio, the numbers are integers with no common factors.

Simplifying integer ratios

Divide each number by a common factor.

$$16 : 12 = 4 : 3$$

$$\div 4$$

Simplifying non-integer ratios

Multiply both numbers to make them both integers.

$$7 : 2.5 = 14 : 5$$

$$\times 2$$

Comparing ratios

To compare two ratios, write them so that both have the same number on one side.

To compare 3 : 1 and 9 : 2, rewrite the first ratio as 9 : 3.

Simplify each ratio. Write it in the correct position in the table.

The first ratio has been done for you.

15 : 10   12 : 28   20 : 5   7 : 21   30 : 18

9 : 27   36 : 24   40 : 24   45 : 105   72 : 18

4 : 1	1 : 3	5 : 3	3 : 2	3 : 7
			15 : 10	

Write each ratio in its simplest form.

a) 10 : 0.5   b) $\frac{1}{4}$: 4   c) $2\frac{1}{2}$: 5   d) 0.1 : 2

....

3 Complete the missing numbers to form equivalent ratios.

a) 16 : .... = 2 : 1   b) 12 : 30 = 2 : ....

c) .... : 56 = 3 : 8   d) 42 : .... = 6 : 5

4 Shami uses red and yellow sweets to decorate a cake in the ratio:

red sweets : yellow sweets = 4 : 3

If Shami uses 20 red sweets, find the number of yellow sweets she uses.

....

5 Nolan makes pink paint by mixing red paint and white paint in the ratio 2 : 5.

a) How much red paint does he mix with 300 ml of white paint?

.................ml

b) Leilani mixes 3 litres of red paint with 7 litres of white paint.

Find which paint is darker, Nolan's or Leilani's. Show your working.

....

6 For each pair of offers, use ratios to show which is better value.

a) Offer 1: a 0.5 kg bag of pasta costs $0.51

Offer 2: a 2 kg bag of pasta costs $2.08

Better value offer: ....

b) Offer 1: a 600 ml carton of milk costs 36 cents.

Offer 2: a 1500 ml carton of milk costs 110 cents.

Better value offer: ....

7 **The ratio of male to female teachers at a school is 2 : 5.**

The number of male teachers is 8.

Sara says, 'The ratio says there are 3 more female teachers than male teachers. So there must be 11 female teachers.'

Do you agree? Explain your answer.

....

....

22.2 Dividing a quantity into a ratio

Summary of key points

An amount can be shared in a ratio.

Example: Share $40 in the ratio 3 : 2.

Number of parts = 3 + 2 = 5

1 part = $40 ÷ 5 = $8

So 3 parts = $24 and 2 parts = $16

Share each amount in the given ratio.

	Amount	Ratio	Answer
a)	$15	2 : 1	$......... and $.........
b)	$32	3 : 1	$......... and $.........
c)	$48	5 : 3	$......... and $.........

2 **In a group of 30 children, the ratio of boys to girls is:**

boys : girls = 2 : 3

Find the number of girls in the group.   ....

3 **Wasim buys 63 plants.**

The plants have either all red flowers or all white flowers.
Wasim buys plants in the ratio:

red : white = 4 : 3

Find the number of plants with red flowers that he buys.   ....

4 **Guy and Hazel share 120 badges between them in the ratio:**

Guy : Hazel = 3 : 5

Find how many badges Hazel gets.   ....

5 **A piece of rope is 1440 cm long.**

It is divided into two pieces in the ratio 2 : 7.

Find the length of each piece.   ....cm and ....cm

6 Theo has a collection of 200 small and large cars.

The ratio of small cars to large cars in his collection is small : large = 4 : 1.

His **small** cars are either red or green in the ratio red : green = 2 : 3.

Find how many small, green cars Theo has.   ....

Think about

7 **Summarise what you know about simplifying ratios and sharing in a given ratio.**

22.3 The unitary method

Summary of key points

In the unitary method you find the value for a single unit or item.

Example: Six identical toothbrushes cost $7.20.

What is the cost of seven of these toothbrushes?

6 toothbrushes   : $7.20

$\downarrow \div 6$ $\downarrow \div 6$

1 toothbrush   : $1.20

$\downarrow \times 7$ $\downarrow \times 7$

7 toothbrushes   : $8.40

Exercise 3

1–4 and 6–7

1 **5 lemons cost 150 cents. Find the cost of:**

a) 1 lemon .... cents   **b)** 3 lemons .... cents

2 **6 bags contain a total of 300 litres of soil.**

Are these statements true or false?

	True	False
a) 12 bags contain a total of 600 litres of soil.	☐	☐
b) To find the amount of soil in 1 bag, divide 6 by 300.	☐	☐
c) 5 bags contain a total of 250 litres of soil.	☐	☐

3 Andrea buys 8 chocolates for $3.20. Work out the cost of:

   a) 10 of these chocolates. $..........

   b) 3 of these chocolates. $..........

4 Prue uses these ingredients to make 25 biscuits.

   500 g flour   250 g butter   150 g sugar

   a) How much flour does she need to make 40 biscuits?

                                                         .... g

   b) Find how much sugar she needs to make 60 biscuits.

                                                         .... g

5 Doug exchanges $80 for Moroccan dirhams. He receives 1000 dirhams.

   a) Jamila exchanges $130 for dirhams. Work out how many dirhams she receives.

                                     .... dirhams

   b) Zeke exchanged some dollars for dirhams. He received 2600 dirhams. Find how many dollars he exchanged.

                                           $..........

6 The working out for a proportion question is:

   $5.60 ÷ 4 = $1.40

   $1.40 × 5 = $7

   Complete the wording of this question so that it matches the working out.

   The cost of four identical items is $5.60. Find ....

   ....

   ....

Think about

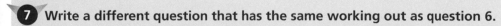

7 Write a different question that has the same working out as question 6.

Probability 2

You will practice how to:

- Design and conduct chance experiments or simulations, using small and large numbers of trials. Analyse the frequency of outcomes to calculate experimental probabilities.

23.1 Probability from experiments

Summary of key points

- A **chance experiment** is an activity where many outcomes are possible, such as rolling a dice.
- We use an **estimate of probability** when it is not possible to use theoretical probability. This estimate of probability is called **relative frequency.**
- The **relative frequency** of a particular outcome is given by the formula

$$\text{Relative frequency} = \frac{\text{number of times that outcome occurred}}{\text{number of trials}}$$

- Estimates for probabilities will vary. The more times an experiment is repeated, the more reliable the estimate of probability.

Exercise 1

1 **For each situation, tick to show if:**

the theoretical probability can be found OR

the probability should be estimated from an experiment.

	Theoretical probability	Experiment
a) The probability of Sam winning a game of chess against a computer.	☐	☐
b) The probability of Sam rolling a 4 on a normal dice.	☐	☐
c) The probability of an egg breaking when Sam drops it from a height of 15 cm.	☐	☐
d) The probability that Sam picks a yellow counter from a bag containing 2 red counters and 1 yellow counter.	☐	☐

2 A game involves throwing a plastic counter. It can land in two different ways.

Standing up
(U)

Lying down
(D)

Sara throws the counter 25 times and records whether it lands up or down.

D	D	D	U	D	D	U	D	U	D
D	U	U	D	D	U	D	D	U	D
U	U	D	D	D					

Tick the outcome you think is more likely.

Standing up ☐   Lying down ☐

Use Sara's results to explain your answer.

....

....

3 Dan works in a bookshop. For 20 people coming into his shop, he records whether they buy something (B) or do not buy anything (D).

| B | B | D | D | D | B | D | D | D | B |
| B | D | D | B | D | D | B | D | B | B |

Use Dan's data to estimate the probability that the next person coming into his shop will buy something.

....

4 Monika throws a biased coin 100 times. The results of her throws are shown in the table.

Result	Frequency
Heads	37
Tails	63

A biased coin is one where each side is not equally likely to occur.

Estimate the probability of throwing heads on the coin.   ....

5 Manuel makes a spinner. He spins his spinner 200 times.

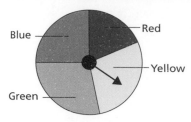

Blue — Red
Yellow
Green

Result	Frequency	Estimate of probability
Red	36	
Yellow	60	
Green	56	
Blue	48	

Complete the final column of the table, giving each value as a decimal.

6 a) Four children each make a spinner.

They test their own spinners, recording the total number of spins and the number of times their spinner lands on red.

Match each child's experiment to the correct relative frequency for red.

Experiments

Aran:
60 spins
24 red

Cai:
72 spins
18 red

Naomi:
50 spins
17 red

Zhen:
80 spins
36 red

Relative frequencies

0.25

0.34

0.4

0.45

Aran = ...   Cai = ...   Naomi = ...   Zhen = ...

b) Fi also makes a spinner. She says:

'I threw my spinner .... times and it landed on red .... times.

The relative frequency that my spinner lands on red is 0.3.'

Complete Fi's statement by writing in possible values.

7 Clara makes a spinner like the one shown. She spins it 40 times. The table shows her results.

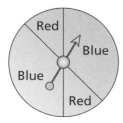

Outcome	Red	Blue
Frequency	9	31

a) Use Clara's results to estimate the probability that the spinner lands on red.

....

b) Explain how Clara could get a more accurate estimate of the probability.

....

....

8 A game involves spinning a plastic plate. It can land in two different ways.

 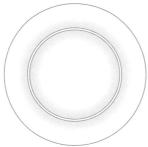

Mi and Ryan each spin the plate 25 times.

Mi's results

Right way up	Upside down
15 times	10 times

Ryan's results

Right way up	Upside down
18 times	7 times

a) Combine Mi's and Ryan's results to estimate the probability of the plate landing upside down when it is spun.

....

b) Mi and Ryan were surprised that they did not get the same results.

Explain why their results are not surprising.

....

....

9 Vehicles can turn left or right at a junction.
Vicki and Sanjay each carry out an independent survey of cars using the junction.
They record how many vehicles turn in each direction.

Vicki's results

Left	Right
103 vehicles	77 vehicles

Sanjay's results

Left	Right
32 vehicles	32 vehicles

a) Work out the experimental probability that a vehicle turns left at the junction using only Vicki's results.

....

b) Work out the experimental probability that a vehicle turns left at the junction using only Sanjay's results.

....

c) Which of these two estimates of experimental probability is likely to be more accurate?

Vicki's ☐   Sanjay's ☐

Explain your answer.

....

....

Think about

10 Design your own spinner. Perform your own experiment to estimate the probability of each outcome.

24 Sequences

You will practice how to:

- Understand term-to-term rules, and generate sequences from numerical and spatial patterns (linear and integers).
- Understand and describe nth term rules algebraically (in the form $n \pm a$, $a \times n$ where a is a whole number).

24.1 Generating sequences

Summary of key points

A **sequence** is a set of numbers, shapes, letters or objects placed in an order that makes a pattern or follows a rule.

Each item in a sequence is called a **term**.

Sequences can be continued by identifying the rule that connects the terms.

The rule that links one term to the next term is called the **term-to-term** rule.

Example: Find the next term in the sequence 4, 7, 10, 13 …

The term-to-term rule is 'add 3'.

The next term in the sequence will be 13 + 3 = 16.

Exercise 1

1 Find the next two terms in each sequence.

a) 34, 29, 24, 19, …….. , ……..

b) −5, 2, 9, 16, …….. , ……..

c) 13, 10, 7, 4, …….. , ……..

d) 4, 25, 46, 67, …….. , ……..

e) 29, 21, 13, 5, …….. , ……..

f) 20, 14, 8, 2, …….. , ……..

g) −37, −30, −23, −16, …….. , ……..

h) 15, 4, −7, −18, …….. , ……..

2 Write the term-to-term rule for each sequence and use it to find the 7th term.

a) 11, 15, 19, 23   Add …..   7th term = ……..

b) 24, 21, 18, 15   Subtract …..   7th term = ……..

c) 31, 26, 21, 16   Subtract …..   7th term = ……..

d) 7, 10, 13, 16   ………………   7th term = ……..

e) 25, 21, 17, 13 .... 7th term = ...

f) −3, 1, 5, 9 .... 7th term = ...

g) 2, 13, 24, 35 .... 7th term = ...

3 Find the required term in these sequences.

	1st term =	Term-to-term rule =	Term to be found
a)	12	Add 5	4th term = ...
b)	22	Subtract 3	4th term = ...
c)	−19	Add 6	6th term = ...

4 Find the missing terms in these sequences.

a) 20, 18, ... , ... , 12   **b)** .... , 16, ... , 26, 31

c) 17, ... , 43, ... , 69

5 Amir and Emily both think of a sequence.

Amir's sequence	Emily's sequence
11, 16, 21, 26...	Term-to-term rule = 'subtract 6'

The sixth term in both sequences is the same.

Work out the second term of Emily's sequence. ....

6 A sequence begins: 27, 45, 63, 81...

Explain how you can easily tell that Angie is wrong.

402 is a number in this sequence.

Angie

....

....

Find the 20th term in these sequences. In each case explain how the 20th term can be found without having to find all the terms in between.

a) 7, 14, 21, 28,

....

....

b) 10, 15, 20, 25,

....

....

Think about

Make up your own sequences which have the following properties: The first term is negative and the third term is 12. Write down the term-to-term rule for each of your sequences.

24.2 Spatial patterns

Summary of key points

Number sequences can be formed from spatial patterns.

Example

Pattern 1   Pattern 2   Pattern 3   Pattern 4

The number of counters in each pattern forms the sequence: 4, 5, 6, 7 …

In each case the number of counters is always 3 more than the pattern number.

Exercise 2

1 Draw the next pattern in each of these sequences and write the number of counters there would be in Pattern 6.

a)
□ □   □ □ □   □ □ □ □
□ □   □ □ □   □ □ □ □
Pattern 1   Pattern 2   Pattern 3   Pattern 4

Pattern 6 = …….. counters

b)

Pattern 1   Pattern 2   Pattern 3   Pattern 4   Pattern 5

Pattern 6 = …….. counters

c)

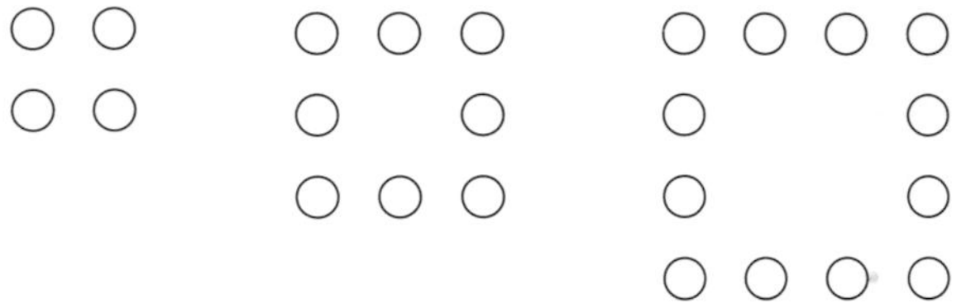

Pattern 1   Pattern 2   Pattern 3   Pattern 4

Pattern 6 = …….. counters

d)

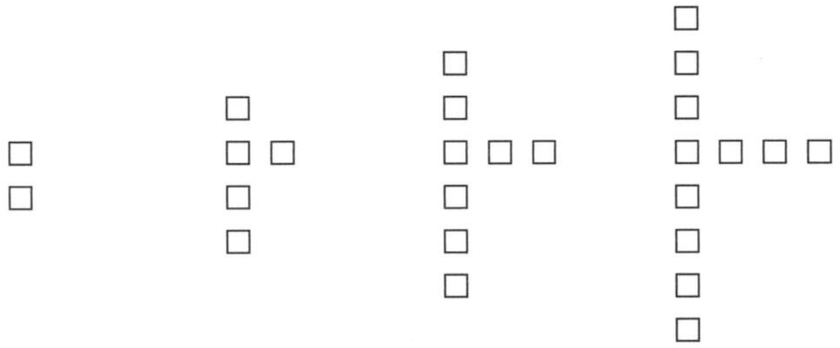

Pattern 1   Pattern 2   Pattern 3   Pattern 4   Pattern 5

Pattern 6 = …….. counters

2 **Here is a pattern made from crosses.**

Pattern 1   Pattern 2   Pattern 3   Pattern 4

a) Complete the table to show the number of crosses in the first six patterns.

Pattern number	1	2	3	4	5	6
Number of crosses	3	6				

b) How many crosses will there be in Pattern 20? ………

c) Explain how you could work out how many crosses there would be in any pattern in this sequence.

………………………………………………………………………………………………

d) Which pattern in the sequence would have 30 crosses?

Pattern ……..

3 Here is a pattern made from circles.

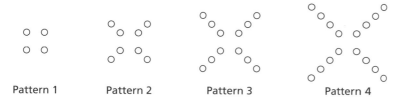

Pattern 1   Pattern 2   Pattern 3   Pattern 4

a) Complete the table to show the number of circles in different patterns.

Pattern number	1	2	3	4	5	...	10	
Number of circles		8	12	16		...		48

b) Write the term-to-term rule for the number of circles.

....

c) Write down how this rule relates to how the pattern is formed.

....

d) Write down how many circles there will be in the 100th pattern. ....

e) Explain how you could work out how many circles there would be in any pattern in this sequence.

....

f) Find which pattern in the sequence has 120 circles. Pattern ....

4 Here are the first three shapes in a pattern.

Pattern 1   Pattern 2   Pattern 3   Pattern 4

a) Draw Pattern 4.

b) Complete the table.

Pattern number	1	2	3	4	5	...	10	
Number of squares		7		9		...		20

c) Find how many squares there are in the 200th pattern. ....

d) Explain how you could work out how many squares there would be in any pattern in this sequence.

....

5 Here is a pattern made from grey and white rectangles.

Pattern 1   Pattern 2   Pattern 3   Pattern 4

a) Complete the table to show the number of grey and white rectangles in each pattern.

Pattern number	1	2	3	4	5	6
Number of grey rectangles	0	1				
Number of white rectangles	2	4				

b) Work out the number of grey and white rectangles in pattern number 20.

Number of grey rectangles = ...   Number of white rectangles = ...

c) Explain how you could work out how many grey rectangles and how many white rectangles there would be in any pattern in this sequence.

Grey

....

White

....

d) One pattern in the sequence has 32 white rectangles. Find how many grey rectangles there are in this pattern.

Think about

6 Make up your own pattern using counters. Write down the number of counters used in each pattern.

24.3 Position-to-term rules

Summary of key points

A **position-to-term** rule is the rule which relates the position number to the value of the term. For a **general term** n, this is called the nth **term rule**.

For example, the nth term rule for this sequence is 3n.

Position number	1	2	3	4	5	6
Term	3	6	9	12	15	18

1 **Write down the first five terms of the sequences with these position-to-term rules:**

   a) the position-to-term rule is add 7.

   ....

   b) the position-to-term rule is multiply by 8.

   ....

   c) the position-to-term rule is subtract 5.

   ....

2 **Each term in a sequence is its position value multiplied by 5.**
   Find the 6th term of the sequence.

   ....

3 **Each term in a sequence is its position value subtract 7.**
   Find the 20th term of the sequence.

   ....

4 **A sequence begins 9, 10, 11, 12, ...**

   Describe how you could find any term in the sequence from its position value.

   ....

5 **The nth term of a sequence is $n + 12$. Find the 15th term of the sequence.**

   ....

6 **The nth term of a sequence is $25n$. Find the 100th term of the sequence.**

   ....

7 **Draw lines to match each position-to-term rule to the correct nth term rule.**

Position-to-term rule	nth term rule
Multiply by 2	$n + 9$
Subtract 2	$9n$
Add 9	$n + 2$
Multiply by 9	$2n$
Add 2	$n - 2$

8 **Write the _n_th term rule for these sequences:**

a)

Position number	1	2	3	4	5	6
Term	4	5	6	7	8	9

_n_th term rule: ....

b)

Position number	1	2	3	4	5	6
Term	5	10	15	20	25	30

_n_th term rule: ....

c)

Position number	1	2	3	4	5	6
Term	−3	−2	−1	0	1	2

_n_th term rule: ....

d)

Position number	1	2	3	4	5	6
Term	13	26	39	52	65	78

_n_th term rule: ....

9 **Here is some information about three sequences.**

Sequence A	Sequence B	Sequence C
Each term is three more than its position value.	Each term is twice its position value.	Each term is one less than its position value.

a) Find the 4th term in sequence A. ....

b) Work out the 35th term in sequence B. ....

c) Find the 100th term in sequence C. ....

d) Find the difference in the 10th terms of sequences B and C. ....

10 **The box contains some information about a sequence.**

Each term in a sequence is 7 times its position value.

Find the sum of the first three terms. ....

11 Richard is making patterns with white and grey square tiles.

Pattern 1　　　　Pattern 2　　　　　　　Pattern 3　　　　　　　Pattern 4

a) Complete the table.

Pattern number	1	2	3	4	5	6
Number of white tiles	3					
Number of grey tiles	3					
Total number of tiles	6					

b) Find the position-to-term rule for the number of white tiles and for the number of grey tiles.

white tiles ....　　　　grey tiles ....

c) For Pattern n, find expressions for the number of

white tiles ....　　　　grey tiles ....

d) Use your answers to part c) to find an expression for the total number of tiles in Pattern n.

Show how you worked out your answer.

....

e) Richard uses 16 grey tiles for one of his patterns.
He says he needs $3 \times 16 = 48$ white tiles for this pattern.

Is Richard correct? Give a reason for your answer.

....

....

Accurate drawing

You will practice how to:

- Draw parallel and perpendicular lines, and quadrilaterals.
- Use knowledge of scaling to interpret maps and plans.
- Visualise and represent front, side and top views of 3D shapes.

25.1 Construction of parallel and perpendicular lines

Summary of key points

Constructing **perpendicular** and **parallel** lines:

A set square can be used:

to draw perpendicular ...

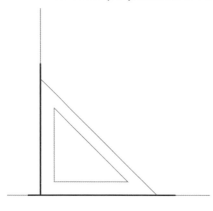

... or parallel lines.

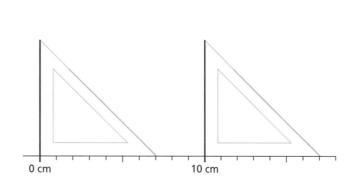

Exercise 1

1 Use a set square to draw a line perpendicular to line *k*.

Line *k*

2 Use a set square to draw a parallel line to line *m*. Make the two lines 2.5 cm apart.

_____ Line *m*

3 Use a set square and a ruler to draw a rectangle measuring 72 mm by 36 mm.

4 Draw this parallelogram accurately.

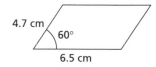

4.7 cm
60°
6.5 cm

5 Construct a trapezium that has:

- Two right angles
- An angle of 50°
- One side measuring 8.5 cm.

Summary of key points

A scale is usually given as a ratio, for example:

- 1 : 25 (1 unit on the drawing represents 25 units in real life)
- 1 : 100 000 (1 cm on the drawing represents 100 000 cm = 1 km in real life).
- 1 cm : 5 m (1 cm on the drawing represents 5 m in real life)

Exercise 2

1. **A map is drawn using the scale 1 cm : 4 m.**

   A building on the map is twice as long as it is wide.

   A fence represented on the map is 20 m long in real life.

   Are these statements true or false?

	True	False
a) In real life, the building is twice as long as it is wide.	☐	☐
b) 4 cm on the map represents 1 m in real life.	☐	☐
c) The fence is 5 cm long on the map.	☐	☐

2. **A map of a park is drawn using a scale of 1 cm : 10 m.**

   a) On the map, a playground is 3 cm wide. How wide is the actual playground? .... m

   b) A path in the park is 120 m long. How long is the path on the map?

   .... cm

3. **Marco makes a scale model of his car using the scale 1 : 10.**

   a) Marco's model is 42 cm long. How long is his car in real life? .... cm

   b) The windscreen of the actual car is 180 cm wide. How wide is the windscreen on the model? .... cm

4 **The diagram shows a scale drawing of a bedroom. The scale is 1 cm : 1 m.**

Key

☐ Window

☐ Bed

a) What are the measurements of the actual bed?

length = ... m   width = ... m

b) The bedroom contains a wardrobe. The actual wardrobe measures 1.2 m by 80 cm. Draw a rectangle to represent the wardrobe on the scale drawing.

5 **A plan of a shopping centre is drawn to a scale of 1 : 200.**

a) On the plan a shop is 3.3 cm wide. What is the width of the actual shop in metres?   .... m

b) A supermarket is 32 m wide. What is the width of the supermarket on the plan?   .... cm

6 **This drawing of three trees uses the scale 1 cm : 5 m.**

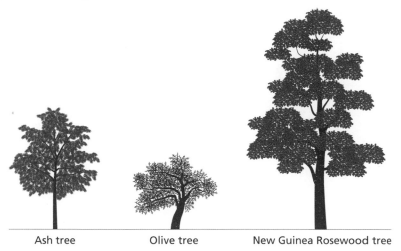

Ash tree   Olive tree   New Guinea Rosewood tree

a) Find the actual height of each tree.

Ash tree .... m

Olive tree .... m

New Guinea Rosewood tree .... m

b) A redwood tree is 95 m in height. It is drawn to the same scale as the other

trees. How tall should the drawing be? .... cm

7 Tina is 1.5 metres tall. The scale drawing shows Tina standing next to an elephant.

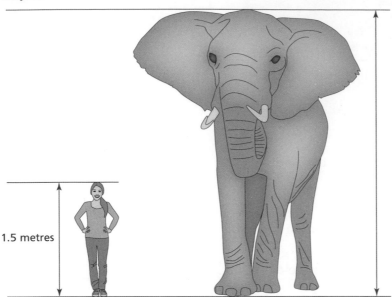

1.5 metres

a) Complete the scale for this diagram. 1 cm : .... m

b) Work out the actual height of the elephant. .... m

8 The diagram shows a scale drawing of a park. The scale is 1 : 15 000.

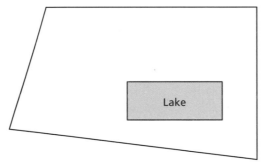

Lake

a) Write down the dimensions of the lake. Give your answer in metres.

.... m by .... m

b) Sal runs once around the perimeter of the park. She says that she runs more than 3 km. Is she correct? Show your working.

....

....

....

....

9 **A map is drawn to a scale of 1 : 500 000.**

a) Find the actual distance that each centimetre on the map represents.
Give your answer in kilometres.

.... km

b) A road measures 7.4 cm on the map. Find the length (in kilometres) of
the actual road.

.... km

c) The distance between two towns is 28 km. How far apart will they be
on the map?

.... cm

10 Theo and Ben each have a map of the same town.

Theo's map has a scale of 1 : 25 000. Ben's map has a scale of 1: 200 000.

A road is 10.4 cm long on Theo's map.

Ben says that same road is 1.3 cm long on his map.

Is Ben correct? Give a reason for your answer.

....

....

....

....

Think about

11 **Find a map of your area on the internet. Use the scale to find the actual distance between some points.**

25.3 Plans and elevations

Summary of key points

A 3D shape can be shown on paper using plans and elevations.

The **plan view** is the view of the shape from the top.

An **elevation** is drawn from the front or the side.

1 Use the centimetre grid. Draw the front and side elevations and the plan view of the cuboid shown below.

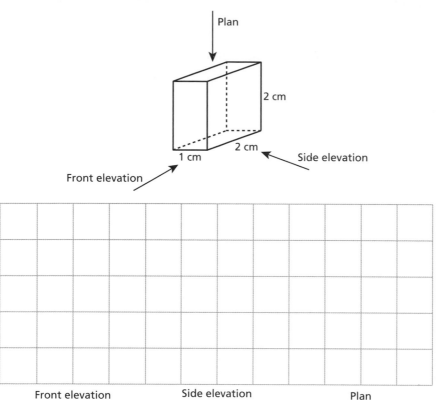

2 Draw the side elevation and the plan for this prism.

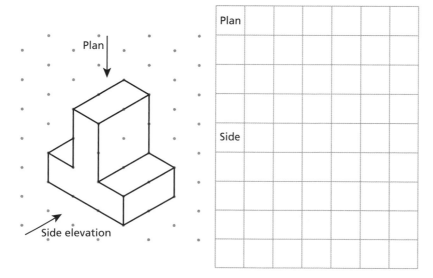

3 The 3D object below is made from six cubes. Draw the plan and the side elevation.

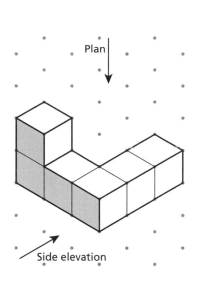

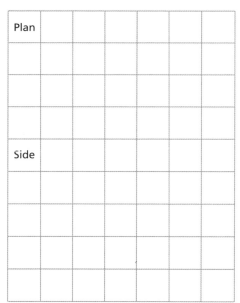

Plan

Side

4 Complete the drawing of the 3D shape and draw the side elevation.

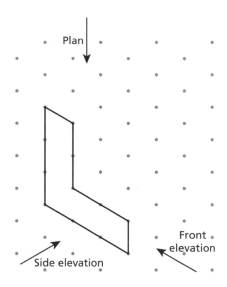

Plan

Front

Side

5 Draw the front elevation and the plan for this square-based pyramid.

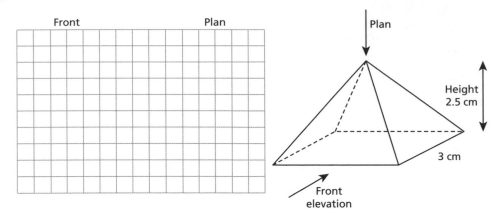

Front   Plan

6 The plan and elevations of a 3D shape made from five cubes is shown below.

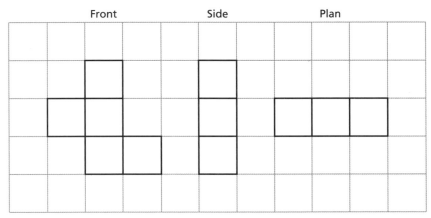

Front   Side   Plan

Draw the shape on the isometric grid.

26 Thinking statistically

You will practice how to:

- Record, organise and represent categorical, discrete and continuous data. Choose and explain which representation to use in a given situation:
  - Venn and Carroll diagrams
  - tally charts, frequency tables and two-way tables
  - dual and compound bar charts
  - waffle diagrams and pie charts
  - frequency diagrams for continuous data
  - line graphs
  - scatter graphs
  - infographics.
- Use knowledge of mode, median, mean and range to describe and summarise large data sets. Choose and explain which one is the most appropriate for the context.
- Interpret data, identifying patterns, within and between data sets, to answer statistical questions. Discuss conclusions, considering the sources of variation, including sampling, and check predictions.

26.1 Choosing appropriate graphs and tables

Summary of key points

To decide which graph or table to use, you need to think carefully about:

- the type of data you are working with
- how you should organise the data you are given
- the reasons why your graph is appropriate.

Exercise 1

1 Here are the marks 20 students scored in a test.

24	12	35	30	18	25	7	14	21	26
26	32	31	19	22	28	42	32	37	9

a) Complete the table, choosing your own groupings. You may not need to use all the rows in the table.

Mark	Tally	Frequency

b) Write a reason why you chose your intervals.

....

c) Draw an appropriate diagram for your table.

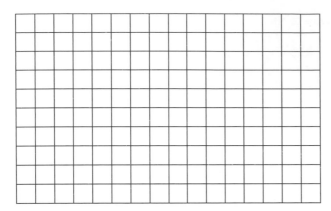

2 Madiha counts how many passengers there are on her bus each morning. The table shows her results.

Number of passengers	Number of mornings
10–14	3
15–19	7
20–24	9
25–29	5
30–34	1

a) Draw an appropriate diagram to show her data.

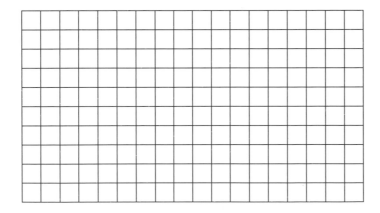

b) Write a reason why you chose your diagram.

....

3 Gordon is concerned with climate change.

He recorded the amount of rain (in mm) that fell in four months in two years.

Here are his results.

	September	October	November	December
Year 1	5	10	15	45
Year 2	10	45	50	95

a) Draw an appropriate diagram to show his data.

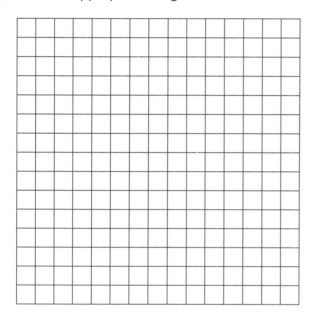

b) Write a reason why you chose your diagram.

....

c) Gordon thinks that there is a change in the weather.

Is he correct? Write a reason for your answer.

....

4 Sarah grows potatoes.

She measured the mass (in grams) and the longest distance (in cm) around each one.

Here is her data.

Mass (g)	37	94	39	67	35	53	60	56
Distance (cm)	12.8	19.6	13.0	17.1	13.2	15.5	16.2	16.4

She thinks that the heavier potatoes have a longer distance around them.

a) Draw an appropriate diagram to show her data.

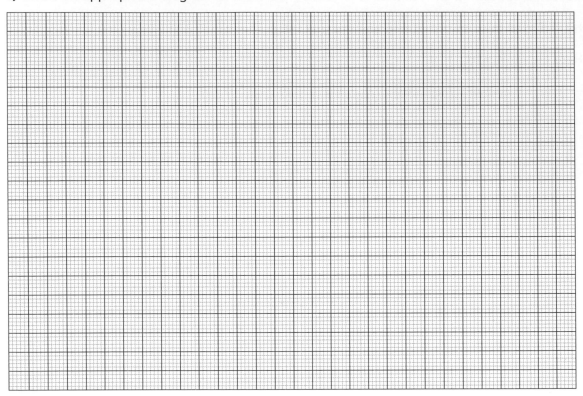

b) Is Sarah correct in what she thinks? Write a reason for your answer.

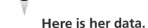

5 **Dagmar asks 120 students in her year group what subject they liked most.**

Here is her data.

Favourite subject	Number of students
Maths	45
History	32
Art	19
Geography	24

She drew the following waffle diagram to show the data she collected.

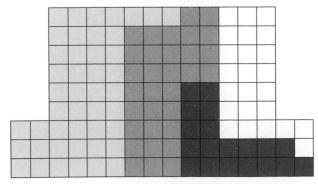

a) Write anything that is wrong or missing from the diagram.

....

....

b) Draw a better waffle diagram to show Dagmar's data.

You do not have to use the whole grid.

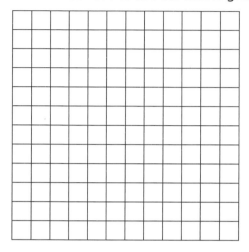

Think about

6 What other types of diagram could Dagmar draw to show her data in question 5?

26.2 Choosing an appropriate average

Summary of key points

To decide which average to use, you need to think carefully about:

- the type of data you are working with
- whether there are any unusually large or small values
- the reasons why your average is appropriate.

Exercise 2

1 State the average that should be used in each case.

a) Marion makes clothing in three sizes: small, medium and large. She wants to know the average clothing size.

....

b) Derrick has some sweets. He gives them to his four children.

Parmida gets 14 sweets.   Raul gets 7 sweets.

Chanya gets 2 sweets.   Waqaar gets 1 sweet.

Waqaar thinks this is unfair and suggests that the number of sweets that each child receives should be the average of these numbers.

....

c) 30 children take a test. The teacher wants to know the mark that half the children scored more than so she works out the average of their marks.

....

2 Zara records the speed of cars passing along a motorway. She wants to know how spread out her data are. What should she calculate?

....

3 Arun records the number of magazines 16 people bought last month.

| 0 | 3 | 4 | 2 | 0 | 1 | 5 | 6 |

| 2 | 0 | 24 | 2 | 4 | 7 | 3 | 1 |

a) Calculate the mean number of magazines bought.

....

b) Find the median number of magazines bought.

....

c) Which of the averages is most appropriate for this data? Explain why.

....

....

4 Raisa counts the number of satsumas there are in each of 20 packets.

Number of satsumas	Frequency
9	3
10	6
11	6
12	4
13	1

a) Calculate an appropriate average for the data.

....

b) Give a reason for your answer to part **a)**.

....

....

5 The table shows the number of sisters that a sample of 50 people have.

Number of sisters	0	1	2	3	4	5	6	7	...	12
Frequency	24	19	3	1	0	2	0	0	...	1

a) Calculate the median number of sisters for this sample.

....

b) Why is the median a better choice of average than the mean for these data?

....

....

27 Relationships and graphs

You will practice how to:

- Understand that a situation can be represented either in words or as a linear function in two variables (of the form $y = x + c$ or $y = mx$), and move between the two representations.
- Use knowledge of coordinate pairs to construct tables of values and plot graphs of linear functions, where y is given explicitly in terms of x ($y = x + c$ or $y = mx$).
- Recognise straight line graphs parallel to the x- or y-axis.
- Read and interpret graphs related to rates of change. Explain why they have a specific shape.

27.1 Linear relationships

Summary of key points

A **linear function** is one which will have a straight line graph when pairs of values that fit the function are plotted on a coordinate grid.

$y = x + 4$ and $s = 3t$ are both examples of linear functions.

Straight line graphs can be used to convert between different amounts, for example between units of measurement or currencies.

Exercise 1

1 **Write the rule for each of these sentences:**

a) a is equal to b multiplied by 5.

....

b) c is equal to four more than d.

....

c) e is equal to six less than f.

....

d) g multiplied by 30 is equal to h.

....

Write a sentence to describe each of these rules:

a) $j = 8k$

....

b) $m = n - 15$

....

c) $p = 12 + q$

....

d) $4r = s$

....

There are 100 cents in every dollar.

a) Write a rule to convert dollars, d, into cents, c.

$$c = \text{...}$$

b) Use your rule to find how many cents there are in 7 dollars.

....

c) Sandy has 1600 cents. How many dollars is this?

....

4 **One Swiss franc, f, is the same as 15 South African rand, r.**

Adam thinks that the rule is $f = 15r$.

Kristi thinks the rule is $r = 15f$.

Who is correct?

....

Explain how you know.

....

....

5 The graph shows the currency conversion between Mexican peso and Japanese yen.

Use the graph to find:

a) The number of yen equivalent to 5 peso.

    ....

b) The number of peso equivalent to 60 yen.

    ....

c) Sam has 80 Mexican peso. Use the graph to help you find how much this is in Japanese yen.

    ....

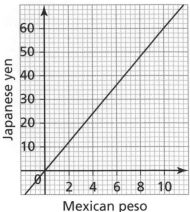

Mexican peso

6 Olga is photocopying test papers.

The photocopier prints 15 sheets each minute.

a) Copy and complete the table to show the number of sheets printed.

Time in minutes (t)	0	1	2	3	4	5
Number of sheets (s)						

b) Draw a graph to show the number of sheets printed with time on the horizontal axis and the number of sheets on the vertical axis.

c) It takes Olga seven minutes to copy the first set of test papers. How many sheets were there?

    ....

d) The second set of test papers has a total of 60 sheets. How long will it take Olga to copy them?

    ....

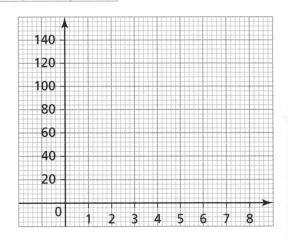

7 One euro is equivalent to four UAE Dirham.

a) Complete the table.

Euro (E)	0	5	10	15	20	25
Dirham (D)						

b) Draw a graph to show the conversion between euro and dirham, with euro on the horizontal axis and dirham on the vertical axis.

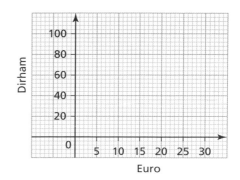

Use your graph to find:

c) the number of dirham equivalent to 14 euro.

....

d) the number of euro equivalent to 84 dirham.

....

e) Serge has 28 euro and Amir has 120 dirham.

Who has more money?

....

Explain how you know.

....

....

8 **In July 2018, 1 British pound (B) was equivalent to 22 Egyptian pounds (E). In October 2019, 1 British pound was equivalent to 17 Egyptian pounds.**

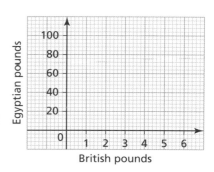

a) Draw graphs to show these two exchange rates.

b) How many more Egyptian pounds was 70 British pounds worth in July 2018 compared to October 2019?

....

27.2 Graphs of linear functions

Summary of key points

The graph of $y = x + 2$ can be drawn by producing a table of values and plotting the points.

x	0	1	2	3
y	2	3	4	5

The graph of $x = 2$ is a vertical line.

The graph of $y = 3$ is a horizontal line.

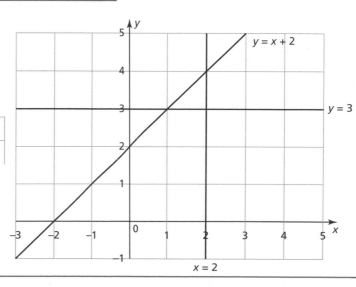

1 These coordinates are all points on the line $y = x + 6$.

Fill in the missing coordinate numbers.

a) (2, ...)   b) (0, ...)   c) (7, ...)   d) (–2, ...)

2 Put a ring around all the points that lie on the line $y = x + 5$.

(0, 5)   (2, 6)   (4, 9)   (6, 10)   (9, 15)

3 Put a ring around all the points that lie on the line $y = 4x$.

(0, 4)   (1, 5)   (3, 12)   (5, 20)   (10, 14)

4 These coordinate points are all on the same straight line except for one.

Circle the coordinate that does not fit the rule.

(3, 9)   (2, 6)   (0, 0)   (1, 4)   (5, 15)   (7, 21)

5 a) Complete the table of values for the line $y = x + 1$.

x	0	1	2	3
y				

b) Complete the table of values for the line $y = 2x$.

x	0	1	2	3
y				

c) Draw the lines $y = x + 1$ and $y = 2x$ on the grid.

d) What are the coordinates of the point where the lines cross?

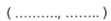
(.........., ...)

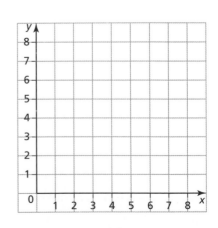

6 a) Complete the table of values for the line $y = x + 3$.

x	0	1	2	3
y				

b) Complete the table of value for the line $y = x$.

x	0	1	2	3
y				

c) Draw the lines $y = x + 3$ and $y = x$ on the grid.

7 Some lines are shown on the grid.

White the equation of each line.

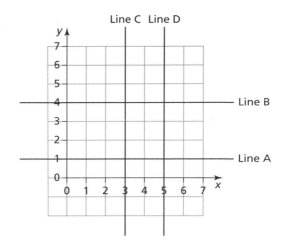

a) Line A   ....

b) Line B   ....

c) Line C   ....

d) Line D   ....

8 Draw each line on the grid and label them with their equation.

a) $x = 1$

b) $y = 7$

c) $x = 6$

d) $y = 5$

e) $y = 0$

f) $x = -3$

g) $y = -2$

9 Put a ring around all the lines that are parallel to the x-axis.

$y = 8$ $y = x$ $y = -1$ $x = 2$

10 Write down the equation of the vertical line that passes through the point (7, 2).

....

11 Write these points in the correct position in the table.

(2, 6)   (6, −2)   (12, 2)   (6, 2)   (4, 2)   (6, 4)

	On $x = 6$	Not on $x = 6$
On $y = 2$		
Not on $y = 2$		

12 Izzah uses horizontal and vertical lines to make a square on a coordinate grid.

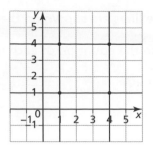

a) What are the lines that she uses?

....

b) She wants to move her square to another position on the coordinate grid, but keep it the same size. Give another possible set of lines that she could use.

....

c) Can you describe how the two vertical lines of the square are linked? What about the two horizontal lines?

....

....

27.3 Graphs in real-life contexts

Summary of key points

Graphs can be used to show some real-life situations, such as a journey.

In a **travel graph**, distance travelled is plotted against time.

A travel graph is sometimes known as a distance–time graph.

Exercise 3

1 The travel graph shows Henrik's journey.

He travels to his friend's house and then back home again.

Complete this description of the journey.

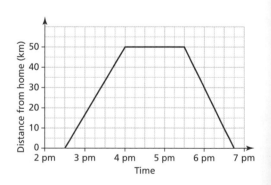

Henrik's friend lives .... km away from Henrik.

Henrik takes .... hours to travel to his friend's house

He leaves his friend's house at ....

and arrives back home again at ....

2 **The travel graph shows a 20 km bus journey.**

a) What time did the bus leave? ....

b) The bus stopped three times on the journey. Work out the total amount of time the bus was stopped.

.... minutes

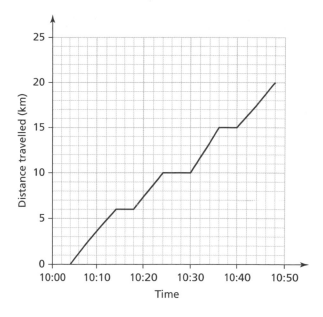

3 **The graph shows the amount of water in a water tank.**

a) How many litres were in the tank at 14:00?

.... litres

Monty takes some water out of the tank.

b) At what time did Monty start taking out some

water? ....

c) How much water did Monty take out of the

tank? .... litres

Some water is put back into the tank.

d) How much water is put back in?

.... litres

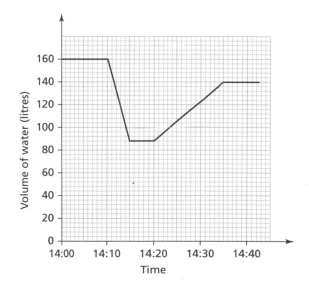

4 Tabina runs herself a bath. The graph shows the depth of water in the bath.

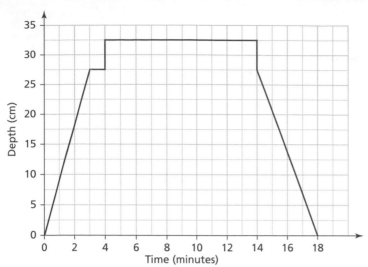

a) How can you tell that Tabina gets in the bath after 4 minutes?

....

....

b) How much does the water level rise when Tabina gets in the bath?

.... cm

c) How long did Tabina spend in the bath?

.... minutes

d) How long does it take for the water to drain from the bath?

.... minutes

5 Draw a travel graph to show Millie's journey from her home to the cinema.

She leaves home at 08:15.

It takes her 10 minutes to drive to the post office, 6 km from her home.

She waits at the post office for 15 minutes, before continuing with her journey.

She reaches the cinema at 09:00. The cinema is 25 km from her home.

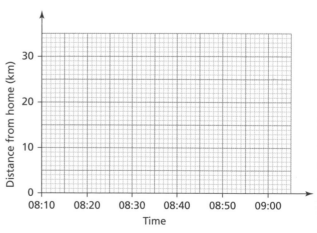